T0233307

Cambridge Elements ≡

Elements in Aerospace Engineering
edited by
Vigor Yang
Georgia Institute of Technology
Wei Shyy
Hong Kong University of Science and Technology

A UNIFIED COMPUTATIONAL FLUID DYNAMICS FRAMEWORK FROM RAREFIED TO CONTINUUM REGIMES

Kun Xu

Hong Kong University of Science and Technology

CAMBRIDGE
UNIVERSITY PRESS

CAMBRIDGE
UNIVERSITY PRESS

University Printing House, Cambridge CB2 8BS, United Kingdom

One Liberty Plaza, 20th Floor, New York, NY 10006, USA

477 Williamstown Road, Port Melbourne, VIC 3207, Australia

314–321, 3rd Floor, Plot 3, Splendor Forum, Jasola District Centre, New Delhi – 110025, India

79 Anson Road, #06–04/06, Singapore 079906

Cambridge University Press is part of the University of Cambridge.

It furthers the University's mission by disseminating knowledge in the pursuit of education, learning, and research at the highest international levels of excellence.

www.cambridge.org
Information on this title: www.cambridge.org/9781108792486
DOI: 10.1017/9781108877534

First published 2021

A catalogue record for this publication is available from the British Library.

ISBN 978-1-108-79248-6 Paperback
ISSN 2631-7850 (online)
ISSN 2631-7842 (print)

A Unified Computational Fluid Dynamics Framework from Rarefied to Continuum Regimes

Elements in Aerospace Engineering

DOI: 10.1017/9781108877534
First published online: May 2021

Kun Xu
Hong Kong University of Science and Technology
Author for correspondence: Kun Xu, makxu@ust.hk

Abstract: This Element presents a unified computational fluid dynamics framework from rarefied to continuum regimes. The framework is based on the direct modelling of flow physics in a discretized space. The mesh size and time step are used as modelling scales in the construction of discretized governing equations. With the variation-of-cell Knudsen number, continuous modelling equations in different regimes have been obtained, and the Boltzmann and Navier-Stokes equations become two limiting equations in the kinetic and hydrodynamic scales. The unified algorithms include the discrete velocity method (DVM)–based unified gas-kinetic scheme (UGKS), the particle-based unified gas-kinetic particle method (UGKP), and the wave and particle–based unified gas-kinetic wave-particle method (UGKWP). The UGKWP is a multi-scale method with the particle for non-equilibrium transport and wave for equilibrium evolution. The particle dynamics in the rarefied regime and the hydrodynamic flow solver in the continuum regime have been unified according to the cell's Knudsen number.

Keywords: multi-scale modelling, non-equilibrium flow computation, gas-kinetic models, Navier-Stokes equations, computational fluid dynamics

ISBNs: 9781108792486 (PB), 9781108877534 (OC)
ISSNs: 2631-7850 (online), 2631–7842 (print)

Contents

1 Introduction

There are two distinguishable modelling scales in gas dynamics: kinetic and hydrodynamic scales. The kinetic scale is the scale of particle mean free path and particle collision time. In this scale, the dynamics of particle transport and collision can be modelled separately. The Boltzmann equation keeps the particle free streaming on the left-hand side and takes account of particle collision on the right-hand side. There should have no more than one collision for an individual particle in the kinetic scale in order to validate the particle free transport modelling on the left-hand side of the Boltzmann equation. In the kinetic scale, the particle moves freely and there is no closed and connected fluid element concept. Numerically, deterministic Boltzmann solvers and a stochastic direct simulation Monte Carlo (DSMC) method have been constructed in such a scale. On the other hand, macroscopic fluid dynamic equations are constructed in a hydrodynamic scale. Even though there is no precise definition of the hydrodynamic scale, the macroscopic equations describe the wave phenomena of accumulating movement of a large number of particles in a quasi-equilibrium state. The hydrodynamic scales must have tens or hundreds of particle mean free paths and particle collision times in order to include enough particles for a well-defined accumulating macroscopic particle behaviour. Navier-Stokes (NS) equations are an example of well-defined hydrodynamic equations. Under the continuum mechanics assumption of no-particle penetration between fluid elements, the NS equations are basically the conservation laws for each fluid element with the viscous stress and heat exchange between the elements. In the regime between the hydrodynamic scale NS and kinetic scale Boltzmann equations, there should be a continuously varying length scale for the description of gas dynamics. However, a variational scale has not been clearly defined between the above two limits for fluid modelling. In this Element, in order to fill up the gap between kinetic and hydrodynamic scale flow modelling, a unified framework according to a cell's Knudsen number, which is defined as the ratio of the particle mean free path over the numerical cell size or the ratio of the particle collision time over the numerical time step, will be constructed for the flow simulation in all regimes. In the continuum regime at a small cell's Knudsen number, the unified algorithm will go to an NS flow solver. In the rarefied regime, it will converge to the Boltzmann solver. The flow physics in the transition regime are determined by the accumulating effect of particle transport and collision within a time step scale for the evolution of both macroscopic flow variables and microscopic gas distribution function. The unified algorithm connects the Boltzmann and NS modelling smoothly with the variation-of-cell Knudsen number.

Due to a continuous scale variation and the difficulty of identifying appropriate flow variables for the description of gas dynamics in the transition regime, no reliable governing equations have ever been proposed. The simplest way is to use brutal force and resolve flow physics down to the particle mean free path scale everywhere. Other attempts try to roughly connect distinguishable NS and Boltzmann solvers through a buffer zone. The flow physics in the buffer zone are assumed to be described correctly by both models. Physically, merging them smoothly is associated with difficulties due to their distinguishable modeling scales. The unified framework is to construct discrete governing equations according to a local cell's Knudsen number and connect different flow regimes seamlessly.

The unified gas-kinetic scheme (UGKS) is an extension of the gas-kinetic scheme (GKS) for the NS solutions. The GKS automatically becomes the limiting scheme of UGKS in the continuum flow regime. As a finite volume flow solver, the GKS only updates macroscopic flow variables through numerical fluxes. Based on the Chapman-Enskog expansion, in the continuum flow regime, the NS gas distribution function is well defined. The GKS uses the integral solution of the kinetic model equation to construct the gas evolution solution from an initial NS distribution function around a cell interface, from which the time accurate numerical flux can be evaluated. In UGKS, in order to capture peculiar non-equilibrium gas distribution function, instead of reconstructing it through the Chapman-Enskog expansion, the gas distribution function is tracked, along with the update of macroscopic flow variables. The same time evolution model is used in UGKS and GKS for the flux evaluation. The integral solution includes the accumulating effect from particle transport and collision within a numerical time step, and it is the key for developing a multi-scale algorithm. Dynamically, it has the similar function as the Riemann solver in establishing CFD for the Euler and NS solutions. The intensity of a particle collision within a time step determines the dynamic equations of the UGKS. In UGKS, both DVM-based gas distribution function and macroscopic flow variables are updated, and the direct modelling algorithm recovers the NS and Boltzmann solution according to the corresponding cell's Knudsen number. Under the UGKS framework, the further extension of the algorithm includes the purely particle-based unified gas-kinetic particle (UGKP) method and the wave-particle-based unified gas-kinetic wave-particle (UGKWP) method. In UGKWP, the gas distribution function is composed of analytic wave and discrete particle, and its evolution is tracked by the same integral solution of the kinetic model equation. According to the cell's Knudsen number $K_{n_c} = \tau/\Delta t$, the number of discrete particles used in the description of the gas distribution function is proportional to $\exp(-1/K_{n_c})$, which varies in different flow regimes. In the highly rarefied hypersonic flow computation, the UGKWP becomes a particle method with notably reduced memory requirements and computational costs in comparison with DVM. In the continuum flow regime, with the diminishing of particles, the UGKWP smoothly returns to the GKS for the update of macroscopic flow variables only, where the analytical wave part of the gas distribution function will automatically converge to the Chapman-Enskog expansion of the NS solution. In the transition regime, the contribution from the discrete particle and analytical wave depends on the local cell Knudsen number. For the low-speed micro-flow, the original DVM-based UGKS still has an advantage in terms of obtaining numerical solutions efficiently, due to the absence of noises from particles. The employ of UGKS or UGKWP for the multiscale flow study depends on the specific engineering applications.

1.1 Multi-scale Modelling for Gas Dynamics

The modelling scale of gas dynamics changes significantly from the deterministic molecular dynamics (MD) by following an individual particle, to the Boltzmann equation by statistically modelling particle transport and collision in the kinetic scale, up to the Navier-Stokes (NS) equations by constructing conservation laws in the hydrodynamic scale. The MD equation captures particle trajectory precisely according to Newton's law, e.g., the particle motion under external field from other particles, and the resolution in the MD goes to the molecular diameter. For the Boltzmann equation, the modelling is up-scaled to the particle mean free path and mean particle collision time. In such a kinetic scale, the information about the precise trajectory of an individual particle gets lost. Instead, the particle is described by probability in each velocity range, e.g., the so-called gas distribution function, which is distributed continuously in space and time through mean-field approximation. In the kinetic scale, the particle will not take more than one collision in order to model particle transport and collision separately, such as the uncoupled particle free streaming and binary collision in the Boltzmann equation. The separate representation of particle transport and collision implicitly enforces the kinetic scale in the Boltzmann modelling. The same constraints on the cell size and time step have been adopted in the DSMC method and many direct Boltzmann solvers because they are based on the same kinetic modeling scale in the construction of numerical algorithms. If a particle takes more than one collision, it will not follow the free streaming movement all the time, and the accumulation of multiple collisions has to be taken into account in the equation. At the current stage, there is no such governing equation with the inclusion of multiple particle collisions. Due to the decoupling of particle transport and collision, it should be emphasized that the Boltzmann equation is only valid in the kinetic scale of particle mean free path and collision time. The solution beyond this scale can be only obtained through the accumulation of a kinetic scale solution, but with the resolution sticking on the kinetic scale, such as the constraints on the cell size and time step in the DSMC simulation. Beyond the particle mean free path and collision time scales, it is difficult to handle a continuously varying scale mathematically to construct the corresponding dynamics, such as modelling the physics for a scale with several particle mean free path resolutions, where the accumulation of multiple particle collisions with zigzag movement can be hardly described mathematically in partial differential equations, even the flow variables cannot be clearly defined.

On the macroscopic level, the Navier-Stokes (NS) equations were constructed by applying Newton's law to the macroscopic fluid element with the implementation of a constitutive relationship and Fourier's law. The hydrodynamic scale for the NS equations, such as the size of a fluid element, has never been precisely defined, which is usually claimed as microscopically large and macroscopically small. It is commonly agreed that the hydrodynamic scale should at least be about one or two orders of magnitude larger than the particle mean free path. The hydrodynamic scale should be large enough to include a very large number of particles within each fluid element, and the mass exchange between neighbouring elements can be ignored. The NS equations describe the wave phenomena from collective particle motion in the quasi-equilibrium state with intensive collisions. In such a scale, each fluid element is assumed to be a closed particle system with the same amount of mass along its movement, and the NS equations provide the conservation laws for each fluid element under force interaction and heat exchange between the elements. Even with volume change along the trajectory of the fluid element, the mass exchange between neighbouring elements is prohibited in the modelling in order to implement the thermodynamic equation of state which is built upon a closed system. The hydrodynamic time scale is the time interval for a wave propagating through an element. Furthermore, the hydrodynamic modelling needs a continuous connection between neighbouring fluid elements, i.e., the so-called continuum mechanics assumption. The Lagrangian formulation of the Euler and NS equations clearly indicates that the connection among neighbouring elements never breaks down. As a result, at the hydrodynamic level, there is actually no non-equilibrium transport mechanism related to the particle penetration. The continuum mechanics assumption for fluid elements in the NS equations, as well as in other extended hydrodynamic equations, will break down gradually with a change of scale from hydrodynamic to kinetic one. Many unsolved problems in fluid mechanics, such as the turbulence, separating flow, and laminar-turbulent transition, may stem from the intrinsic continuum mechanics assumption in the NS formulation, where the non-equilibrium transport mechanism and the abundant discontinuities within the fluid are absent in the mathematical formulation. The breaking down of connected fluid elements and the associated non-equilibrium transport may become important for the description of turbulent flow, such as the rapid emerging of a large amount of degrees of freedom in the laminar turbulence transition. But, the NS-based turbulence modelling has only a quasi-equilibrium transport mechanism with a diffusive process, even with enhanced turbulence viscosity. In the mathematical formulation, there is not much difference between NS gas dynamics equations and equation for the elastic body deformation. There may have intrinsic weakness in the hydrodynamic modelling for the non-equilibrium turbulence flow.

The physical modelling among MD, Boltzmann, and NS equations is associated with a reduction of information in a flow description through a coarse-graining process. Theoretically, the gas evolution can be fully captured through MD formulation without going to the Boltzmann and NS equations at all. However, in reality, it is not practical and possible to follow a large number of particles in the MD resolution, and it is legitimate to find the most efficient and appropriate way to describe flow dynamics in the corresponding regime. For example, for a re-entry problem, the dynamics provided by the Boltzmann equation through the DSMC are accurate, and the method is efficient for the flight simulation above 80 kilometres. At an altitude lower than 40 kilometres, the NS equations can be applied faithfully. Therefore, efficient and distinguishable governing equations are preferred in different flow regimes. The current study is to find out a method which can produce an accurate and efficient description in all flow regimes with adaptive physical modelling. In practice, for hypersonic flow, one single flight may be associated with the existence of multiple flow regimes, such as the compressible high density flow at leading edge and the rarefied one at trailing edge. The development of a unified algorithm is to solve this kind of multi-scale problem.

In order to develop a CFD method for both rarefied and continuum flow simulation, the flow modelling in both kinetic and hydrodynamic scales has to be bridged dynamically with the variation of scale, such that both the Boltzmann and NS equations should become the limiting equations automatically. In the kinetic scale, there are mainly two kinds of numerical methods to solve the Boltzmann equation: the stochastic particle method and the deterministic kinetic solver. Both methods are widely used in academic research and engineering applications. The stochastic method employs discrete particles to simulate the statistical behaviour of molecular gas dynamics. The direct simulation Monte Carlo (DSMC) method is the most successful particle simulation method for rarefied flow (Bird 1994; Oran et al. 1998). The consistency of the DSMC method and the Boltzmann equation has been established mathematically (Wagner 1992). Similar to many direct Boltzmann solvers, the mesh size and time step in the DSMC method are constrained by the particle mean free path and mean collision time (Alexander et al. 1998). Otherwise, due to the operator splitting treatment of particle transport and collision, the numerical dissipative coefficient being proportional to the time step Δt would take over the physical one, which is on the order of the particle collision time τ. On the other hand, since the required number of simulation particles is independent of the Mach and Knudsen numbers, the stochastic particle method seems to use the best adaptive technique in the discretization of particle velocity space. The DSMC method requires low computational memory to cover the whole velocity space and gains high efficiency in rarefied flow computation, especially for the multidimensional high-speed flow. As a particle method, the DSMC suffers from statistical noise due to an insufficient number of particles. It becomes very difficult to simulate low-speed flow and capture small temperature variation. Moreover, with the increase of flow density, such as flight simulation at an altitude below 80 kilometres, the computational cost in 3D DSMC calculation grows dramatically and nonlinearly. The DSMC's intensive particle collision and small cell size requirement make it impossible to simulate flow in the continuum and near continuum regime. In order to address the stiffness problem in the continuum regime, asymptotic preserving (AP) Monte Carlo methods (Pareschi & Russo 2000; Ren et al. 2014) were developed so that the Euler solution can be obtained in the hydrodynamic limit without constraints on the time step and mesh size, as required by the traditional DSMC method. Unfortunately, these AP schemes can only use the Euler limit instead of an NS one in the continuum flow regime. As a result, the accuracy of AP-DSMC in the near continuum regime, when the cell size and time step are larger than the particle mean free path and collision time, is questionable. The stochastic particle methods based on kinetic model equations, such as the Bhatnagar-Gross-Krook (BGK), the ellipsoidal statistical BGK (ES-BGK) models (Gallis & Torczynski 2000; Macrossan 2001; Tumuklu et al 2016; Fei et al. 2020), and the Fokker-Planck (FP) model (Jenny et al. 2010; Gorji & Jenny 2015), have been constructed for further improvement of computational efficiency. In order to recover the NS solution in the continuum flow regime, instead of AP property, the schemes have to satisfy the unified preserving (UP) property (Guo et al. 2020), where the coupled discretization of particle transport and collision within a time step becomes necessary.

The deterministic methods employ a discrete distribution function to solve the kinetic equations. The discrete velocity methods (DVM) for Boltzmann and kinetic model equations have been extensively studied in the last several decades (Chu 1965; Yang & Huang 1995; Aristov 2012; Tcheremissine 2005; Li & Zhang 2004; Wu et al. 2015), which have great advantages for simulating low-speed microflow (Huang et al. 2013; Wu et al. 2014). In order to improve the efficiency and remove time-step limitations on the deterministic kinetic solvers, AP schemes (Larsen et al. 1987; Jin 1999) and kinetic-fluid hybrid methods (Degond et al. 2010) have been proposed and constructed. Accurate solutions can be obtained without statistical noises, and high efficiency can be achieved using numerical acceleration techniques, such as implicit (Mieussens 2000; Zhu et al. 2017a, 2019a), multigrid (Zhu et al. 2017b), high-order/low-order (HOLO) decomposition (Chacon et al. 2017); memory reduction techniques (Chen et al. 2017); a fast spectral method for the full Boltzmann collision term (Mouhot & Pareschi 2006; Wu et al. 2013); and an adaptive refinement method (Chen et al. 2012). For both stochastic and deterministic methods, once the gas evolution process is split into collisionless particle transport and instant collision, a numerical dissipation that is proportional to the time step will be introduced implicitly. Therefore, the mesh size and the time step in many kinetic solvers need to be less than the particle mean free path and collision time. The analysis of the dissipative mechanism in the kinetic flux vector splitting scheme and the Lattice Boltzmann method (LBM) will be presented in several places in this Element.

Besides the methods starting from the kinetic equation and being extended to the continuum regime, other approaches are coming from the NS side. Based on the hydrodynamic formulation, the extended hydrodynamics and moment equations developed in the continuum flow regime are pushed to the transition and rarefied flow. Based on the Chapman and Enskog expansion, the Euler, Navier-Stokes, Burnett, and super-Burnett equations can be derived (Chapman et al. 1990). In shock-wave calculation, the Burnett and super-Burnett equations are subject to numerical instabilities, and the equations need to be regularized (Zhong et al. 1993). Another interesting approach is the moment method (Grad 1949), where the distribution function is expanded in terms of a complete set of orthogonal polynomials, e.g., the Hermite polynomials, with the coefficients corresponding to the velocity moments. This expansion is truncated up to a certain order, resulting in a closed set of moment equations. Although the Grad moment approach has been widely used, it is best suited for problems where the velocity distribution function can be expressed as a perturbation of the equilibrium state, i.e., the near continuum flow. Recently, the Chapman-Enskog expansion and the Grad moment methods have been combined in the development of regularized methods (Levermore 1996), such as the well-defined R13 and R26 (Struchtrup 2005; Struchtrup & Torrihon 2003; Gu & Emerson 2009). The success of these methods in low-speed microflow with a modest Knudsen number has been confirmed. For high-speed flow, on the basis of Eu's generalized hydrodynamics (Eu 2016), the balanced closure has been recently developed (Myong 2001) and the hypersonic rarefied flow simulations have been presented (Jiang et al. 2019b). The uncertainty in the extended hydrodynamic and moment equations is the absence of any clear definition of modelling scale in the construction of these equations. They are basically derived from equation to equation, not from the direct modelling in different scales. The number of flow variables that is suitable for a valid description of non-equilibrium flow is also unclear. Intrinsically, all macroscopic flow variables are space-averaged quantities on the hydrodynamic scale, which is far beyond kinetic scale modelling. Theoretically, these equations lack the non-equilibrium transport mechanism due to the underlying continuum mechanics assumption, where the no-penetration and unbreakable connection between neighbouring fluid elements are intrinsically rooted in these models. As a result, these equations can cope with the diffusive and dispersive process for the description of quasi-equilibrium state evolution. In fact, it becomes hard to capture the non-equilibrium transport, even though the dissipative coefficients can be enlarged and have a delicate dependence on the flow variables. In the continuum flow regime, the intensive particle collisions merge an individual particle's movement into a collective 'wave' behaviour, and their dynamic evolution can be described by a few flow variables, such as the mass, momentum, and energy in the NS equations. With the increase of rarefaction, the individual particle's contribution to the flow dynamics becomes important, where the closed fluid element assumption breaks down. In a highly rarefied flow, individual particle movement becomes gradually independent with free penetration. The flow physics from continuum to rarefied is associated with a dramatic increase of information with the change of modelling scale, which can be hardly described by a few predefined macroscopic flow variables. Extending this modelling process continuously from hydrodynamic to kinetic scale is difficult due to the following reasons. First, it is unclear how to define a continuous variation of modelling scale between the kinetic and hydrodynamic one and to get the corresponding equations under such a varying scale. Second, it is unclear what kind of flow variables are appropriate for describing the flow physics between these two limits. Third, there is no clear scale separation in the transition regime. Conventional mathematical tools, such as asymptotic expansion, may not be applicable to cover the whole regime. In the NS equations, there are only five flow variables. For the Boltzmann equation, the particles are basically independent, and there is a theoretically infinite number of degrees of freedom for capturing the gas distribution function. Between the hydrodynamic and kinetic limits, it is essentially unclear how many flow variables and what kinds of variables are appropriate in the moments and extended hydrodynamic equations. But, these questions need to be properly answered in the construction of a unified algorithm. The existing extended and irreversible thermodynamics mainly focus on the study of the flow close to equilibrium. There is not much knowledge available regarding how to describe non-equilibrium properties through a macroscopic thermodynamic formulation. The flow physics between the kinetic and hydrodynamic scales have not been properly explored theoretically. A direct task for the unified algorithm is to clearly define a modelling scale and construct the gas dynamics in such a scale.

Other popular numerical algorithms for both continuum and rarefied flow simulations are the hybrid methods. The hybrid methods, such as the combination of the NS and the DSMC or NS and direct Boltzmann solver (Schwartzentruber & Boyd 2007; Degond et al. 2006; Tiwari 1998; Wijesinghe & Hadjiconstantinou 2004), have been used in engineering applications. Since the Boltzmann solver is much more expensive than the NS solver, the NS solver should be applied in a domain as large as possible. In most hybrid methods, a buffer zone is designed to couple different approaches with the exchange of information from different solvers. These methods may depend sensitively on the location of the interface, where many criteria have been proposed to define the buffer zone with the requirement of applicability of both methods in this region. Two well-known hybrid methods are the CFD/DSMC and CFD/Boltzmann solver (Schwartzentruber et al. 2008; Burt et al. 2011). For example, it is commonly stated that in the buffer zone, the flow can be described by both NS equations and kinetic solvers. In fact, this is exactly where the difficulty arises with applying the DSMC in the NS regime. For example, in order to avoid numerical dissipation from the DSMC solution, the cell size in the buffer zone has to be less than the particle mean free path and the buffer zone needs to be located in the kinetic regime. On the other hand, once the cell size in the buffer zone is on the particle mean free path scale, the assumptions in modelling NS are no longer valid. For the DSMC/CFD hybrid method, another difficulty concerns how to overcome the statistical fluctuation in DSMC. A CFD solver may be sensitive to the noise introduced through the boundary and become numerically unstable. Another hybrid method is the combination of the CFD and Boltzmann solver, where a gas-kinetic scheme (GKS) is used as a CFD solver (Kolobov et al. 2007). Recently, in order to improve the efficiency of numerical computation, a hybrid approach is formulated by combining the multi-scale unified gas-kinetic scheme (UGKS) and GKS (Xiao et al. 2020), and the success of this hybrid method is from the multi-scale nature of the UGKS, which is valid in any flow regime but can be replaced by a more efficient method in the continuum flow regime. Distinguished from the above hybrid method, a general synthetic iterative scheme (GSIS) has been developed in recent years. The ingredient of GSIS is that macroscopic synthetic equations are simultaneously solved with gas kinetic equations, from which the macroscopic flow properties are obtained to guide the evolution of a gas molecular system. Due to the direct use of macroscopic governing equations, the GSIS asymptotically preserves the Navier-Stokes equations at the continuum limit, so that the restriction on spatial grid cell size is eliminated. (Su et al. 2020; Zhu et al. 2021).

1.2 Unified Gas-Kinetic Schemes

For the high-speed flow around a vehicle in near space flight, flow physics of different regimes may appear simultaneously in a single computation, such as the highly non-equilibrium shock layer, high density leading edge, low density trailing edge, and wake flow. Figure 1 presents the local Knudsen number around a vehicle at Mach number 4 and Reynolds number 59,373, which is calculated by the DVM-based unified gas-kinetic scheme (UGKS). As shown in this figure, the local Knudsen number can cover a wide range of values, from 10^{-5} to 10 with six orders of magnitude difference. A single governing equation, such as the Boltzmann or NS equations, can hardly be applied efficiently in all regions to obtain a reliable physical solution. A unified algorithm to cover all regimes from the compressible NS solution to the free molecular flow is necessary. In fact, a large portion of the flow around the vehicle stays in the transition regime, which connects hydrodynamic and kinetic ones smoothly. In this calculation, it is impossible to use a cell size that is less than the particle mean free path everywhere, and it is unlikely to define a buffer zone to bridge

$M_\infty = 4$, Re = 59373

Minimum Kn_{GLL} is: 7.61e–005

Maximum Kn_{GLL} is: 1.99e+001

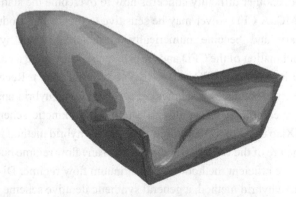

Figure 1 A local Knudsen number around a vehicle at Mach number 4 and Reynolds number 59,373. Courtesy of D.W. Jiang (Jiang et al. 2019a).

the DSMC and the NS solver. Therefore, a continuous variation of governing equations from the Boltzmann to the NS has to be modelled and used in order to obtain a solution efficiently and accurately. In other words, an efficient algorithm should be able to model the flow dynamics under a variation of the local Knudsen number. A continuous scale-dependent governing equation is not purely an averaging of the Boltzmann equation over a large scale. Instead of using ensemble averaging on the kinetic equation, the dynamics on a large scale has to be modelled directly, while an evolution solution with the inclusion of particle transport and collision in the numerical time step or cell size scales can be used in the construction of the unified algorithm. The unified algorithm establishes governing equations with a variable ratio of the cell size over the local particle mean free path, and is able to connect the Boltzmann and NS dynamics smoothly.

In this Element, we are going to present a general framework to model and compute multi-scale non-equilibrium flow. The basis of the unified algorithm is to use the cell size Δx and time step Δt as the physical modelling scales in the construction of gas dynamic equations. The cell Knudsen number, which is defined by $Kn_c = \ell/\Delta x$ or $Kn_c = \tau/\Delta t$ with the local particle mean free path ℓ, the particle collision time τ, the numerical cell size Δx, and the time step Δt, becomes a physical parameter to model the flow physics in the discretized space directly. Based on the physical conservation laws, both macroscopic flow variables and microscopic gas distribution function will be updated in the discretized space, and the local flow evolution depends on the accumulating solution from the particle transport (ℓ) and collision (τ) up to the modelling scales ($\Delta x, \Delta t$). Under the direct modelling framework, a unified gas-kinetic scheme (UGKS) has been developed for all flow regimes (Xu & Huang 2010; Huang et al. 2013; Liu et al. 2016; Liu & Xu 2017). The UGKS will recover the NS solution ($Kn_c \leq 10^{-3}$) and Boltzmann transport ($Kn_c \geq 1$) in the limiting cases, and provide a smooth solution in the whole transition regime ($10^{-3} \leq Kn_c \leq 1$). The multi-scale flow physics used in the design of the UGKS comes from an explicit time-dependent evolving solution of a kinetic model equation. This evolution solution takes into account the effect of multiple particle collisions in a numerical time step, from which the flux transport across a cell interface is obtained in the determination of the updates of macroscopic flow variables and gas distribution function in a finite control volume. The evolution solution with coupled particle transport and collision plays a key role in the development of a unified algorithm, and the solution is a function of the local cell Knudsen number. The UGKS achieves high efficiency in the continuum flow regime with the use of the same mesh size as the standard NS solver, which can be much larger than the particle mean free path. The scheme becomes the Boltzmann solver when the time step is close to the local particle collision time. A smooth transition between these two limits is obtained. Similar to the other DVM, the UGKS uses mesh points to discretize the particle velocity space in order to update the gas distribution function. For high-speed flow computation, the scheme is expensive, with a huge memory and computational cost stemming from the use of a very large number of grid points in the particle velocity space. In order to improve the efficiency of the UGKS, a particle version of the unified algorithm, which is referred to as a unified gas-kinetic particle (UGKP) method, is developed (Liu et al. 2020; Zhu et al. 2019b). Instead of covering the whole velocity space with discrete velocity points, the UGKP uses particles to track the evolution of the gas distribution function under the same UGKS framework. In UGKP, the time step can be much larger than the particle collision time, and the particle's collision effect will be taken into account within each time step. More specifically, the particle in UGKP is categorized as a collisional and collisonless particle within a time step. The collisional particle will be eliminated in the cell, and its contribution to the macroscopic conservative flow variables will be recorded; then it will be re-sampled from the updated flow variables of all colliding particles at the beginning of the next time step. In UGKP, with the update of macroscopic flow variables, the equilibrium state can be quantitatively evaluated and is used to guide the evolution of a discrete particle. So, the dynamical evolution of colliding particles in UGKP, even though they are eliminated within a time step, is captured analytically through the evolution of macroscopic flow variables. Therefore, the time step in UGKP can be larger than the particle collision time, and the evolution solution can be captured accurately for collisional and collisionless particles. In the multi-scale computation, there is no constraint on the cell size and time step in the UGKP, because multiple collisions for individual particles have been modelled on the scale of a time step. In order to further improve the efficiency of the UGKP, a wave-particle formulation is introduced in the construction of a unified gas-kinetic wave-particle (UGKWP) method (Liu et al. 2020; Zhu et al. 2019b). In the UGKWP, only the free transport particle in the next time step evolution needs to be re-sampled from the corresponding macroscopic flow variables at the beginning of the next time step. The dynamic evolution of the collisional particle, such as the contribution to the flux evaluation, can be evaluated analytically. In other words, only the non-equilibrium part in the gas distribution function will be captured by discrete particle and the equilibrium part is modelled by following the evolution of macroscopic flow variables analytically. As a result, in the flow simulation, the number of particles tracked in the UGKWP is proportional to $e^{(-\Delta t/\tau)}(\rho/m) = e^{(-1/Kn_c)}(\rho/m)$, where ρ is density and m is the molecular mass. Therefore, the computational efficiency can be greatly improved in the continuum flow regime, and the statistical noises from the discrete particle are effectively eliminated due to the absence of particles. In the highly rarefied flow regime ($Kn_c \geq 1$), the UGKWP becomes the particle method, which is the most efficient one for hypersonic flow computation. For the continuum flow simulation, the gas-kinetic scheme (GKS) for the NS solution has been well-developed in the past decades (Xu 2001; Xu 2015) and the UGKWP will exactly recover GKS in the continuum limit, which has similar computational costs and memory requirements as a standard NS solver. The UGKWP models the evolution of the gas distribution function with a wave-particle decomposition with adaptive weights between them, according to the cell Knudsen number. Therefore, in the UGKWP, the flow description can be smoothly switched from the hydrodynamic wave to the kinetic particle and connect these two distinguishable gas dynamics modeling. Furthermore, due to the particle nature of the UGKWP in the highly rarefied flow regime, the single collision time kinetic relaxation model that is used in the construction of the scheme can be directly modified according to the particle mean free path and velocity, such as ($\tau \sim \ell/|\mathbf{u}|$). This modification only changes the particle's travelling distance and does not affect the conservation of the scheme. The positive effect of this justification is clearly observed in the shock structure calculation by the UGKWP. Based on the wave-particle decomposition, the kinetic model equation can be further developed through direct modelling.

In the following sections, we will first introduce the gas-kinetic scheme (GKS) as a kinetic equation–based NS solver, which is the basis for the development of a unified scheme. Then, the basic methodology of direct modelling will be introduced and the unified gas-kinetic scheme will be presented. In order to improve the efficiency of UGKS in hypersonic flow simulation, the unified gas-kinetic particle method is shown, which is followed by the introduction of unified gas-kinetic wave particle method. The UGKWP is probably the most efficient method in CFD for the high-speed rarefied and continuum flow simulation. It combines the advantages of the particle method for the highly rarefied flow and the deterministic method for the continuum flow. The UGKWP goes to the GKS precisely in the continuum flow limit. The use of waves and particles in the modelling of fluid dynamics abandons the continuum mechanics assumption used in the construction of macroscopic fluid dynamic equations, and provides a general framework to describe both equilibrium and non-equilibrium transport. The particle penetration between fluid elements is automatically incorporated into the algorithm. In other words, the unified framework presents a physical description of gas dynamics with an intrinsic multi-scale nature. The flow physics of different regimes are obtained under a single framework with the variation-of-cell Knudsen number. Besides gas dynamics, the methodology of UGKS has been successfully applied for simulating several other non-equilibrium transports, such as radiative and neutron transfer, plasma, and multiphase dispersive flows. It is expected that the unified framework will also provide a useful tool in the turbulence study through the modelling of freely moving and penetrating fluid elements in order to capture the non-equilibrium transport rather than using the purely diffusive process or quasi-equilibrium mechanism in the current turbulent modelling.

2 The Gas-Kinetic Scheme

2.1 Governing Equations

The Boltzmann equation for a single-component monatomic gas without external force is (Chapman et al. 1990)

$$\frac{\partial f}{\partial t} + \mathbf{u} \cdot \nabla_{\mathbf{x}} f = Q(f,f), \tag{2.1}$$

where $\mathbf{x} = (x,y,z)^t \in \mathcal{R}^3$ is the space variable, $\mathbf{u} = (u,v,w)^t \in \mathcal{R}^3$ is the particle velocity, and $f(\mathbf{x},t,\mathbf{u})$ is the velocity distribution function. The binary collisions are described by the collision term $Q(f,f)$,

$$Q(f,f) = \int_{\mathcal{R}^3} \int_{\mathcal{S}^2} (f'_* f' - f_* f) |\mathbf{u_r}| \sigma d\Omega d\mathbf{u}_*.$$

where the shorthand notation $f'_* = f(\mathbf{x},t,\mathbf{u}'_*)$ is used, similarly for f' and f_*. Here $\mathbf{u}, \mathbf{u}_*$ are the pre-collision particle velocities and $\mathbf{u}', \mathbf{u}'_*$ are the corresponding post-collision velocities. Ω is a unit vector in $\mathcal{S}^2$ along the relative post-collision velocity $\mathbf{u}' - \mathbf{u}'_*$. The differential cross section σ measures the probability of collision, which depends on the strength of relative velocity and deflection angle between pre-collision and post-collision velocities. In the provided Boltzmann equation, the left-hand side is the free streaming for the particle transport and the right-hand side is the collision term. The transport and collision in the Boltzmann equations are decoupled in the modelling. In order to keep the free streaming valid, each particle cannot suffer more than one collision and the modelling scale in the equation should be on the particle mean free path and collision time.

From the Boltzmann equation, the H-theorem is proved for the system to approach an equilibrium state which is uniquely defined for a spatially homogeneous state, i.e., the Maxwellian distribution function,

$$g = \rho \left(\frac{m}{2\pi k_B T} \right)^{3/2} \exp \left(-\frac{m}{2k_B T} (\mathbf{u} - \mathbf{U})^2 \right), \tag{2.2}$$

where ρ is the density, T is the temperature, $\mathbf{U} = (U, V, W)$ is the macroscopic velocity, m is the molecular mass, and k_B is the Boltzmann constant. The macroscopic quantities can be obtained by taking moments of the velocity distribution function f,

$$\mathbf{W} = \begin{pmatrix} \rho \\ \rho \mathbf{U} \\ \rho E \end{pmatrix} = \int \psi f d\mathbf{u}, \tag{2.3}$$

where $\mathbf{W}$ are the densities of conservative flow variables, such as mass ρ, momentum $\rho\mathbf{U}$, and energy ρE; $\psi = (1, \mathbf{u}, \mathbf{u}^2/2)^T$ is the vector for conservative moments; and $d\mathbf{u} = dudvdw$.

The microscopic Boltzmann equation can be used to derive the macroscopic hydrodynamic equations through the Chapman-Enskog theory. With the definition of Knudsen number $Kn = \ell/L$ as the ratio of particle mean free path over the characteristic length scale, when the Knudsen number is small, the Euler equations,

$$\frac{\partial \rho}{\partial t} + \nabla_{\mathbf{x}} \cdot (\rho \mathbf{U}) = 0,$$

$$\frac{\partial (\rho \mathbf{U})}{\partial t} + \nabla_{\mathbf{x}} \cdot (\rho \mathbf{U} \mathbf{U} + p\mathbf{I}) = 0, \tag{2.4}$$

$$\frac{\partial (\rho E)}{\partial t} + \nabla_{\mathbf{x}} \cdot ((\rho E + p)\mathbf{U}) = 0,$$

can be obtained from the zeroth-order asymptotic expansion of the Boltzmann equation, and the first-order asymptotic expansion of the Boltzmann equation gives the Navier-Stokes equations,

$$\frac{\partial \rho}{\partial t} + \nabla_{\mathbf{x}} \cdot (\rho \mathbf{U}) = 0,$$

$$\frac{\partial (\rho \mathbf{U})}{\partial t} + \nabla_{\mathbf{x}} \cdot (\rho \mathbf{U} \mathbf{U} + p\mathbf{I}) = \nabla_{\mathbf{x}} \cdot (\mu \sigma(\mathbf{U})), \tag{2.5}$$

$$\frac{\partial (\rho E)}{\partial t} + \nabla_{\mathbf{x}} \cdot ((\rho E + p)\mathbf{U}) = \nabla_{\mathbf{x}} \cdot (\mu \sigma(\mathbf{U}) \cdot \mathbf{U} + \kappa \nabla_{\mathbf{x}} T),$$

where $p = \int c^2 f d\mathbf{u}/3$ is pressure, $\sigma(\mathbf{U})$ is the strain rate tensor,

$$\sigma(\mathbf{U}) = (\nabla_{\mathbf{x}} \mathbf{U} + (\nabla_{\mathbf{x}} \mathbf{U})^T) - \frac{2}{3} \mathrm{di} \, v_{\mathbf{x}} \mathbf{U} \mathbf{I},$$

κ is the thermal conductivity coefficient, and μ is the dynamic viscosity coefficient. The dynamic viscosity coefficient for the hard sphere or variable hard sphere molecule has the form

$$\mu = \mu_{ref}\left(\frac{T}{T_{ref}}\right)^{\omega},$$ (2.6)

where μ_{ref} and T_{ref} are the reference viscosity and temperature and ω is the temperature dependency index. For a variable soft sphere molecule, the viscosity is related to the mean free path in the equilibrium state (Bird 1994),

$$\lambda = \frac{4\alpha(7 - 2\omega)(5 - 2\omega)}{5(\alpha + 1)(\alpha + 2)}\left(\frac{m}{2\pi k_B T}\right)^{1/2}\frac{\mu}{\rho}.$$ (2.7)

For the hard sphere or variable hard sphere model, $\alpha = 1$.

The Boltzmann equation describes the time evolution of a gas distribution function in the kinetic scale. The provided macroscopic NS and Euler equations are the equations in the hydrodynamic scale, which is much larger than the kinetic one. In the hydrodynamic scale, due to the inclusion of intensive particle collision, as in cases with a local small Knudsen number, the gas distribution function gets close to the local equilibrium state. At the same time, the flow description changes from particle free transport and collision to accumulating wave propagation. In the hydrodynamic scale, the flow evolution is described by a significantly reduced number of flow variables, such as the mass $\rho(\mathbf{x}, t)$, momentum $\rho\mathbf{U}(\mathbf{x}, t)$, and energy $\rho E(\mathbf{x}, t)$. The derivation of hydrodynamic equations from the Chapman-Enskog expansion is based on the following two assumptions (Chapman et al. 1990):

1. The gas distribution function is close to the local equilibrium state, such as $f = g + f^{(1)} + f^{(2)} + \ldots$, and $f^{(1)}, f^{(2)}$ are proportional to the small Knudsen number Kn and Kn^2.

2. The normal solution of the distribution function f depends on the space and time $(\mathbf{x}, t)$ through macroscopic flow variables, such as $f(\mathbf{x}, t, \mathbf{u}) = f(\rho(\mathbf{x}, t), \rho\mathbf{U}(\mathbf{x}, t), \rho E(\mathbf{x}, t), \mathbf{u})$.

The function of the first assumption is that the deviation from the equilibrium state is small, which can be used to linearize the Boltzmann collision term, such as

$$ff_1 \simeq (g + f^{(1)})(g_1 + f_1^{(1)}) \simeq gg_1 + f^{(1)}g_1 + f_1^{(1)}g,$$

where g_1 and g are the same equilibrium state. The second assumption is to separate the hydrodynamic and kinetic scales in the Boltzmann equation. The spatial and temporal variations of f (i.e., f_t and f_x on the left-hand side of the Boltzmann equation) must go through macroscopic flow variables, varying on the hydrodynamic scale, to determine its evolution. Therefore, the assumption enforces the left-hand side of the Boltzmann equation (transport part in the kinetic scale) to follow macroscopic flow motion in the hydrodynamic scale. Since the collision term in the Boltzmann equation affects the kinetic scale only, the separation of the scales leads the Boltzmann equation to be solved in a successive Chapman-Enskog approximation. Here, the characteristic length scale for the hydrodynamic part is defined as the length scale for the variation of macroscopic flow variables, such as $L = \rho/|\partial\rho/\partial x|$. The provided two assumptions underlying the Chapman-Enskog expansion cannot be valid in the transition regime, with a continuous variation of scale from kinetic to the hydrodynamic one. The Chapman-Enskog method does not include the scale variation effect. The further expansion, which is supposed to cover equations in the transition regime, such as Burnett and super-Burnett, requires the evolution of f through macroscopic flow variables in the hydrodynamic scale as well. In reality, in the transition regime, there is no physical basis for f to depend on $(\mathbf{x}, t)$ through macroscopic variables. The Chapman-Enskog expansion does not fill up the gap between NS and Boltzmann modelling. Another point that needs to be clarified concerns the validity of the Boltzmann equation in all flow regimes. This conclusion is actually based on resolving the kinetic scale flow physics everywhere. The Boltzmann equation itself does not directly describe the flow evolution beyond the kinetic scale. The solution in other scale is from the time accumulation of the Boltzmann solution in the kinetic scale. Besides the distinguishable modelling scales of the Boltzmann and NS equations, the starting point of the unified algorithm is to recognize the scale variation and use the mesh size and time step as modelling scales directly in the construction of the gas dynamic equations. This scale can connect smoothly the kinetic mean free path to the hydrodynamic fluid element scale. The mesh size and time step provide a physical reality which can be used to model the flow physics and present the physical solution. In other words, in the unified algorithm, there is no gap between kinetic and hydrodynamic scale modelling.

Besides the full Boltzmann equation, the kinetic model equations are also very useful in the study of non-equilibrium flow. The relaxation Bhatnagar-Gross-Krook (BGK) equation is highly successful in this way (Bhatnagar et al. 1954):

$$\frac{\partial f}{\partial t} + \mathbf{u} \cdot \nabla_{\mathbf{x}} f = (g - f)/\tau, \tag{2.8}$$

where τ is the particle collision time and is related to the viscosity coefficient $\tau = \mu/p$. The provided BGK model gives a unit Prandtl number in the corresponding NS equations, which can be fixed through the modification of the heat flux or the viscous stress. In this section, based on the BGK model of a monatomic gas and Prandtl number fix, the gas kinetic scheme (GKS) for the NS solution will be introduced. For diatomic gas, a similar scheme can be constructed with equilibrium and non-equilibrium internal energy distribution (Xu 2001; Xu et al. 2005; Xu et al. 2008).

The aim of this Element is to present a unified framework for the development of a numerical algorithm which connects the Boltzmann and NS equations and provides an accurate solution for the whole transition regime. The GKS presented next is basically an NS solver (Xu 2001), which uses macroscopic variables to construct the initial gas distribution function and follows its evolution at a cell interface for the flux evaluation. The GKS has a similar computational cost as the standard NS solver. The importance of GKS is that it will become a limiting flow solver in the continuum flow regime for the multi-scale methods, such as the UGKS, UGKP, and UGKWP methods presented in later sections (Liu et al. 2020; Xu & Huang 2010; Zhu et al. 2019b). In the continuum flow regime, the cell size used in the computation is usually much larger than the particle mean free path. The massive amount of particle collisions in the cell size and time step scales will present a very small cell Knudsen number $Kn_C = \ell/\Delta x$, and the multi-scale methods will converge to a hydrodynamic flow solver with the dynamic formation of a near-equilibrium distribution function. For the NS solution, the distribution function is well-defined by the Chapman-Enskog expansion, which will be used in the construction of GKS, and it will be recovered automatically from UGKS without explicitly using the Chapman-Enskog expansion. Since the NS gas distribution function in the continuum flow regime can be constructed from macroscopic flow variables, the GKS is essentially a macroscopic flow solver.

Over the past 20 years, the GKS for Navier-Stokes solutions has been developed and successfully applied from nearly incompressible to hypersonic viscous and heat-conducting flow computations. The GKS is a finite volume method for the update of macroscopic flow variables only with the flux functions evaluated from a time-dependent gas distribution function at the cell interface. Here a continuous particle velocity space is used in the gas distribution function. The macroscopic flow variables, including the fluxes, are the moments of the analytic distribution function. The efficiency of the GKS is similar to the Riemann solver-based methods, where the same Courant-Friedrichs-Lewy (CFL) condition is used in the determination of the time step. The GKS is able to present an accurate NS solution in the smooth flow region once the solution is well-resolved by the mesh size, such as the boundary layer, and has favourable shock-capturing capability in the non-equilibrium shock region due to the implementation of a particle free transport mechanism in providing necessary and delicate numerical dissipation. Even for the continuum flow simulation, it seems that keeping a non-equilibrium mechanism is still very helpful in the construction of a stable numerical shock structure. More importantly, the unified algorithms presented in later sections will get back to GKS in the continuum flow limit. This property of going back to an NS solver directly is paramountly important for any claimed multi-scale method. Otherwise, the multi-scale method would have an intrinsic weakness if it could not go back to the GKS or other standard NS solvers in the hydrodynamic limit. To automatically recover an NS solver from the multi-scale method seems more appropriate than the direct enforcement of the macroscopic governing equations through the so-called hybrid-type approximation between kinetic and hydrodynamic solvers explicitly. Understanding GKS is helpful for fully appreciating the modelling in the unified algorithm.

2.2 Gas-Kinetic Scheme for the Navier-Stokes Solutions

For the continuum flow computation, the GKS is a finite volume scheme for the update of cell-averaged conservative flow variables. Due to the conservation in the particle collisional process inside each control volume, there is no source term in the update of conservative flow variables inside each cell, and an accurate evaluation of the interface flux determines the quality of the scheme. In GKS, starting from the initial reconstruction of macroscopic flow variables, a time-dependent gas distribution function will be constructed for the flux evaluation. The coupling of particle transport and collision in the gas evolution process at a cell interface is the key to the success of GKS, and it is also the main difference between GKS and many other kinetic solvers, such as the kinetic flux vector splitting (KFVS) scheme (Pullin 1980; Deshpande 1986), Lattice Boltzmann method (LBM; Chen & Doolen 1998), and many discrete velocity methods (DVM; Chu 1965; Mieussens 2000; Li & Zhang 2004).

Similar to the MUSCL type approach (van Leer 1979), the first step of GKS is to interpolate the macroscopic flow variables inside each computational cell. For a second-order GKS, the van Leer limiter is typically used for the initial data reconstruction based on the conservative, primitive, or characteristic variables (van Leer 1977). Based on the reconstructed initial data around each cell interface, the corresponding gas distribution function as an initial condition can be determined. Then, a time-dependent gas distribution function f will be obtained from the integral solution of the kinetic model equation, from which the numerical flux is evaluated by taking moments of the distribution function. Besides incorporating the gradients of flow variables in the normal direction on both sides of a cell interface, the gradients in the tangential directions will also be included in the time evolution of the gas distribution function, i.e., the so-called multidimensional property (Xu et al. 2005). In the following discussion, a three-dimensional GKS will be presented for the flux evaluation across a plane with the normal x-direction and two tangential y- and z-directions. Under the general geometry with arbitrary normal direction, the GKS can be constructed similarly with coordinate transformation (Xu 2015), even with unstructured mesh (Pan & Xu 2016).

Denote $(x = 0, y = 0, z = 0)$ as the centre of the cell interface with the normal x-direction. Note that for a high-order method, multiple Gaussian points may be used at the cell interface. On both sides of this interface, the macroscopic variables are reconstructed as

$$\mathbf{W} = \begin{cases} \mathbf{W}^l(0,0,0) + \nabla\mathbf{W}^l \cdot \mathbf{x}, & x < 0, \\ \mathbf{W}^r(0,0,0) + \nabla\mathbf{W}^r \cdot \mathbf{x}, & x \geq 0, \end{cases} \tag{2.9}$$

where $\mathbf{x} = (x, y, z)$ and $\mathbf{W} = (\rho, \rho U, \rho V, \rho W, \rho E)^T$ are the densities of mass, momentum, and energy in the three-dimensional case. For the continuum flow computation, the GKS updates the conservative flow variables inside each control volume $(\Omega_{\mathbf{x}} = \Omega_{i,j,k})$ in the physical space in the following way:

$$\mathbf{W}_{i,j,k}^{n+1} = \mathbf{W}_{i,j,k}^n - \frac{1}{\Omega_{\mathbf{x}}} \sum_s \int_{t^n}^{t^{n+1}} \mathbf{F}_s \cdot \mathbf{n}_s S dt = \mathbf{W}_{i,j,k}^n - \frac{1}{\Omega_{\mathbf{x}}} \sum_s \mathbf{F}_{\mathbf{W}}, \tag{2.10}$$

where $\mathbf{F}_s$ is the flux at the S_{th} interface surface with area S in the out-normal direction $\mathbf{n}$. The flux $\mathbf{F}_s$ is evaluated from a time-dependent gas distribution function. The above equation is basically a direct application of physical conservation law in a discretized space. It is better not to regard the above equation as the conservative moments of the Boltzmann equation on a physical control volume. As mentioned earlier, the Boltzmann equation is constructed on the kinetic scale, but the above cell size is on the hydrodynamic scale. If Eq. (2.10) were considered as a moment equation of the Boltzmann equation, the fluxes $\mathbf{F}_s$ in the above equation would be automatically evaluated from the particle free transport due to the free transport modelling on the left-hand side of the Boltzmann equation. This is a common mistake made by most kinetic solvers. The above equations are the direct modelling of physical conservation laws in a control volume Ω_x. As a macroscopic flow solver, the scales of cell size and time step in Eq. (2.10) are much larger than the kinetic particle mean free path and collision time. Therefore, the time evolution of the gas distribution function at a cell interface will take into account the massive particle collisions within a time step. The flow physics at the scale of $\Delta t \gg \tau$ is definitely not the particle free transport in the interface flux evaluation. Theoretically, the conservation laws in Eq. (2.10) are valid in any scale of the cell size and the corresponding time step under the CFL condition, and in any flow regime, even in the free molecular one. But the flux transport across the cell interface depends on the ratio of the numerical time step Δt to the particle collision time τ, in which the flow physics is directly modelled in a discretized space.

For the gas kinetic scheme, a time-dependent gas distribution function f at the central point of the cell interface, i.e., $(x_{i+1/2}, y_j, z_k) = (0, 0, 0)$ and time t, can be obtained based on the integral solution of BGK model,

$$f(x_{i+1/2}, y_j, z_k, t, u, v, w) = \frac{1}{\tau} \int_0^t g(x', y', z', t', u, v, w) e^{-(t-t')/\tau} dt'$$
$$+ e^{-t/\tau} f_0(x_{i+1/2} - ut, y_j - vt, z_k - wt), \qquad (2.11)$$

where $x' = x_{i+1/2} - u(t - t'), y' = y_j - v(t - t'), z' = z_k - w(t - t')$ are the trajectory of a particle motion and f_0 is the initial gas distribution function f at the beginning of each time step *(t = 0)*. The terms related to f_0 and g represent different scale gas dynamics, and the solution gives an evolution process from the kinetic particle free transport of f_0 to the hydrodynamic equilibrium state evolution of g. This scale-dependent transport solution makes the development of a multi-scale method possible. Any successful scheme for multi-scale transport must intrinsically include a similar mechanism for non-equilibrium to equilibrium gas evolution through a time-dependent process, with the inclusion of both kinetic and hydrodynamic flow physics. Both g and f_0 have to be fully determined in Eq. (2.11) in order to obtain a time-dependent solution f. In the kinetic model, the distribution function f and the equilibrium g have the same mass, momentum, and energy at any point of space and time, which satisfies the compatibility condition,

$$\int f(\mathbf{x}, t, \mathbf{u}) \psi_\alpha d\mathbf{u} = \int g(\mathbf{x}, t, \mathbf{u}) \psi_\alpha d\mathbf{u}, \alpha = 1, 2, 3, 4, 5, \qquad (2.12)$$

where $\psi = (1, u, v, w, \frac{1}{2}(u^2 + v^2 + w^2))^T$ and $d\mathbf{u} = du dv dw$.

For a multidimensional GKS, the initial gas distribution function f_0 is constructed as

$$f_0 = \begin{cases} g^l\Big(1 + a^l x + b^l y + c^l z - \tau(a^l u + b^l v + c^l w + A^l)\Big) & x \leq 0 \\ g^r\Big(1 + a^r x + b^r y + c^r z - \tau(a^r u + b^r v + c^r w + A^r)\Big), & x < 0 \end{cases}$$

$$(2.13)$$

where g^l and g^r are Maxwellian distribution functions on the left- and right-hand sides of a cell interface, such as

$$g = \rho\left(\frac{\lambda}{\pi}\right)^{3/2} \exp\left[-\lambda\big((u - U)^2 + (v - V)^2 + (w - W)^2\big)\right],$$

where $\lambda = m/(2kT)$, a^l, and a^r are related to the slopes in the expansion of a Maxwellian in the normal direction, b^l and b^r are the slopes in one tangential direction, and c^l and c^r are in another orthogonal tangential direction. The τ related terms in Eq. (2.13) come from the Chapman-Enskog expansion for the non-equilibrium part of the NS solution. Note that the non-equilibrium parts have no net contribution to conservative flow variables, i.e.,

$$\begin{cases} \int (a^l u + b^l v + c^l w + A^l)\psi_\alpha g^l d\mathbf{u} = 0, \\ \int (a^r u + b^r v + c^r + A^r)\psi_\alpha g^r d\mathbf{u} = 0. \end{cases}$$

$$(2.14)$$

The above compatibility conditions are equivalent to the Euler equations for the inviscid flow.

All terms in f_0, such as $g^l, g^r, a^l, a^r, b^l, b^r, c^l$, and c^r, can be fully determined from the reconstructions of macroscopic flow variables, which will be given later.

After constructing f_0, the equilibrium state g around $(x = 0, y = 0, z = 0, t = 0)$ is modelled as

$$g = g_0\Big(1 + (1 - \mathrm{H}[x])\bar{a}^l x + \mathrm{H}[x]\bar{a}^r x + \bar{b}y + \bar{c}z + \bar{A}t\Big),$$

$$(2.15)$$

where $\bar{a}^l$ and $\bar{a}^r$ are the slopes in the normal direction on both sides of the cell interface, $\bar{b}$ and $\bar{c}$ are the common slopes in the tangential directions, and $\mathrm{H}[x]$ is the Heaviside function defined by

$$\mathrm{H}[x] = \begin{cases} 0, & x < 0, \\ 1, & x \geq 0. \end{cases}$$

Here g_0 is a local Maxwellian distribution function located at the centre of a cell interface $(x = 0, y = 0, z = 0)$. In both f_0 and g, all coefficients, such as $a^l, b^l, ..., \overline{A}$, come from the derivatives of a Maxwellian distribution function in space and time.

The dependence of $a^l, b^l, ..., \overline{A}$ on the particle velocity can be obtained from the Taylor expansion of a Maxwellian, with the following form:

$$a^l = a_1^l + a_2^l u + a_3^l v + a_4^l w + a_5^l \frac{1}{2}(u^2 + v^2 + w^2) = a_\alpha^l \psi_\alpha,$$

$$A^l = A_1^l + A_2^l u + A_3^l v + A_4^l w + A_5^l \frac{1}{2}(u^2 + v^2 + w^2) = A_\alpha^l \psi_\alpha,$$

$$\cdots$$

$$\overline{A} = \overline{A}_1 + \overline{A}_2 u + \overline{A}_3 v + \overline{A}_4 w + \overline{A}_5 \frac{1}{2}(u^2 + v^2 + w^2) = \overline{A}_\alpha \psi_\alpha,$$

where $\alpha = 1, 2, 3, 4, 5$ and all coefficients $a_1^l, a_2^l, ..., \overline{A}_5$ are local constants.

With the initial data reconstruction in Eq. (2.9), the macroscopic flow variables $\mathbf{W}$ on the left- and right-hand sides of a cell interface are provided. The connection between the gas distribution function f and the macroscopic variables is

$$\int g^l \psi_\alpha d\mathbf{u} = \mathbf{W}^l(0,0,0); \quad \int g^l a^l \psi_\alpha d\mathbf{u} = \mathbf{n} \cdot \nabla \mathbf{W}^l \qquad (2.16)$$

$$\int g^r \psi_\alpha d\mathbf{u} = \mathbf{W}^r(0,0,0); \quad \int g^r a^r \psi_\alpha d\mathbf{u} = \mathbf{n} \cdot \nabla \mathbf{W}^r, \qquad (2.17)$$

where $\nabla \mathbf{W}^l$ and $\nabla \mathbf{W}^r$ are the gradients of macroscopic variables on the left- and right-hand sides of a cell interface, and $\mathbf{n}$ is the normal direction. Similarly, in the tangential directions $\mathbf{t}_1$ and $\mathbf{t}_2$, b^l, b^r, c^l and c^r can be obtained from

$$\int g^l b^l \psi_\alpha d\mathbf{u} = \mathbf{t}_1 \cdot \nabla \mathbf{W}^l; \quad \int g^r b^r \psi_\alpha d\mathbf{u} = \mathbf{t}_1 \cdot \nabla \mathbf{W}^r, \qquad (2.18)$$

$$\int g^l c^l \psi_\alpha d\mathbf{u} = \mathbf{t}_2 \cdot \nabla \mathbf{W}^l; \quad \int g^r c^r \psi_\alpha d\mathbf{u} = \mathbf{t}_2 \cdot \nabla \mathbf{W}^r. \qquad (2.19)$$

With the definition of Maxwellian distributions in 3D,

$$g^l = \rho^l \left(\frac{\lambda^l}{\pi}\right)^{\frac{3}{2}} e^{-\lambda^l((u-U^l)^2 + (v-V^l)^2 + (w-W^l)^2)}$$

and

$$g^r = \rho^r \left(\frac{\lambda^r}{\pi}\right)^{\frac{3}{2}} e^{-\lambda^r((u-U^r)^2 + (v-V^r)^2 + (w-W^r)^2)},$$

from Eq. (2.16) and (2.17), the equilibrium states g^l and g^r are fully determined by the density, velocity, and temperature, which can be evaluated from the local reconstructed conservative flow variables. For example, with the total energy density

$$\rho E = \frac{1}{2}\rho\left(U^2 + V^2 + W^2 + \frac{3}{2\lambda}\right),$$

λ in the Maxwellian can be obtained. The evaluation of the moments of the equilibrium state is presented in the Appendix.

The parameters a^l and a^r are related to the local slopes of macroscopic flow variables. For example, the solution for the coefficients in the parameters a^l and a^r can be found in

$$M_{\alpha\beta}a_\beta = M(a_1, a_2, a_3, a_4, a_5)^T = \frac{1}{\rho}\mathbf{n}\cdot\nabla\mathbf{W},$$

on the left- and right-hand sides separately, where the matrix M is given by

$$M_{\alpha\beta} = (1/\rho)\int g\psi_\alpha\psi_\beta d\mathbf{u}.$$

The solutions for the above equations are given in the Appendix. Similarly, the parameters b^l, b^r, c^l, and c^r can be determined from Eq. (2.18) and (2.19).

After determining the terms a^l, b^l, c^l, a^r, b^r, and c^r from the initially reconstructed macroscopic flow variables, A^l and A^r of f_0 in Eq. (2.13) can be found from Eq. (2.14), which are

$$M^l_{\alpha\beta}A^l_\beta = -\frac{1}{\rho^l}\int \psi_\alpha(a^l u + b^l v + c^l w)g^l d\mathbf{u},$$

$$M^r_{\alpha\beta}A^r_\beta = -\frac{1}{\rho^r}\int \psi_\alpha(a^r u + b^r v + c^r w)g^r d\mathbf{u},$$

where $M^{l,r}_{\alpha\beta} = \int g^{l,r}\psi_\alpha\psi_\beta d\mathbf{u}/\rho^{l,r}$. The solutions of A^l and A^r can be obtained using the formula in the Appendix.

For the equilibrium state g in Eq. (2.15), the corresponding values of ρ_0, U_0, V_0, W_0, and λ_0 in g_0,

$$g_0 = \rho_0\left(\frac{\lambda_0}{\pi}\right)^{\frac{3}{2}}e^{-\lambda_0((u-U_0)^2+(v-V_0)^2+(w-W_0)^2)}, \tag{2.20}$$

can be found as follows. Taking the limit $t \to 0$ in Eq. (2.11) and substituting its solution into Eq. (2.12), the conservation constraint at the centre of the interface and $t = 0$ gives

$$\int g_0 \psi_\alpha d\mathbf{u} = \mathbf{W}_0 = \int_{u>0} \int g^l \psi_\alpha d\mathbf{u} + \int_{u<0} \int g^r \psi_\alpha d\mathbf{u}, \qquad (2.21)$$

where $\mathbf{W}_0 = (\rho_0, (\rho U)_0, (\rho V)_0, (\rho W)_0, (\rho E)_0)^T$. The modelling in the above equation is that the equilibrium state at the cell interface at $t = 0$ is determined from the colliding particles from both sides according to particle velocity, and this equilibrium state provides the direction the local system will evolve to. Since g^l and g^r were obtained previously, the above moments can be evaluated explicitly; see the Appendix for the moments evaluation. Then, g_0 is uniquely determined. For example, λ_0 in g_0 can be found from

$$\lambda_0 = 3\rho_0 / \left[4\left((\rho E)_0 - \frac{1}{2}((\rho U)_0^2 + (\rho V)_0^2 + (\rho W)_0^2)/\rho_0 \right) \right].$$

Then, $\bar{a}^l$ and $\bar{a}^r$ in the expansion of g (see Eq. (2.15)) can be obtained through the relation of

$$\frac{\mathbf{W}_{i+1}(\Delta x/2, 0, 0) - \mathbf{W}_0}{\rho_0 \Delta x^+} = \overline{M}^0_{\alpha\beta} \begin{pmatrix} \bar{a}^r_1 \\ \bar{a}^r_2 \\ \bar{a}^r_3 \\ \bar{a}^r_4 \\ \bar{a}^r_5 \end{pmatrix} = \overline{M}^0_{\alpha\beta} \bar{a}^r_\beta, \qquad (2.22)$$

and

$$\frac{\mathbf{W}_0 - \mathbf{W}_i(-\Delta x/2, 0, 0)}{\rho_0 \Delta x^-} = \overline{M}^0_{\alpha\beta} \begin{pmatrix} \bar{a}^l_1 \\ \bar{a}^l_2 \\ \bar{a}^l_3 \\ \bar{a}^l_4 \\ \bar{a}^l_5 \end{pmatrix} = \overline{M}^0_{\alpha\beta} \bar{a}^l_\beta, \qquad (2.23)$$

where Δx^- and Δx^+ are the distances between the cell interface to the cell centres in the normal direction. Since the matrix $\overline{M}^0_{\alpha\beta} = \int g_0 \psi_\alpha \psi_\beta d\mathbf{u}/\rho_0$ is known, $(\bar{a}^r_1, \bar{a}^r_2, \bar{a}^r_3, \bar{a}^r_4, \bar{a}^r_5)^T$ and $(\bar{a}^l_1, \bar{a}^l_2, \bar{a}^l_3, \bar{a}^l_4, \bar{a}^l_5)^T$ can be evaluated accordingly. The term $\bar{b}$ and $\bar{c}$ in Eq. (2.15) can be evaluated by

$$\int \bar{b} g_0 \psi d\mathbf{u} = \int_{u>0} b^l g^l \psi d\mathbf{u} + \int_{u<0} b^r g^r \psi d\mathbf{u}, \qquad (2.24)$$

$$\int \bar{c} g_0 \psi d\mathbf{u} = \int_{u>0} c^l g^l \psi d\mathbf{u} + \int_{u<0} c^r g^r \psi d\mathbf{u}, \qquad (2.25)$$

or from reconstructed macroscopic flow variables in the tangential directions of the cell interface.

Up to this point, all parameters in the initial gas distribution function f_0 and the equilibrium state g at the beginning of each time step $t = 0$ are determined. After substituting Eq. (2.13) and Eq. (2.15) into Eq. (2.11), the gas distribution function f at a cell interface can be expressed as

$$
\begin{aligned}
f(0,0,0,t,u,v,w) = {} & C_1 g_0 \\
& + C_2\Big(\bar{a}^l u \mathrm{H}[u] + \bar{a}^r u(1 - \mathrm{H}[u]) + \bar{b}v + \bar{c}w\Big)g_0 \\
& + C_3 \bar{A} g_0 \\
& + C_4\Big(\mathrm{H}[u]g^l + (1 - \mathrm{H}[u])g^r\Big) \\
& + C_5\Big((ua^l + vb^l + wc^l)\mathrm{H}[u]g^l \\
& \quad + (ua^r + vb^r + wc^r)(1 - \mathrm{H}[u])g^r\Big) \\
& + C_6\Big((ua^l + vb^l + wc^l + A^l)\mathrm{H}[u]g^l\Big) \\
& + C_6\Big((ua^r + vb^r + wc^r + A^r)(1 - \mathrm{H}[u])g^r\Big)
\end{aligned}
\tag{2.26}
$$

where
$$
\begin{aligned}
C_1 &= 1 - e^{-t/\tau}, \\
C_2 &= te^{-t/\tau} - \tau(1 - e^{-t/\tau}), \\
C_3 &= t - \tau(1 - e^{-t/\tau}), \\
C_4 &= e^{-t/\tau}, \\
C_5 &= -te^{-t/\tau}, \\
C_6 &= -\tau e^{-t/\tau}.
\end{aligned}
\tag{2.27}
$$

The only unknown left in the above expression is $\bar{A}$. Since both f (Eq. (2.26)) and g (Eq. (2.15)) contain $\bar{A}$, the integration of the conservation constraint Eq. (2.12) at *(0,0,0)* over the whole time step Δt gives

$$
\int_0^{\Delta t} \int (g - f)\psi_\alpha dt\, d\mathbf{u} = 0,
$$

which reduces to

$$
\begin{aligned}
\overline{M}^0_{\alpha\beta}\bar{A}_\beta = {} & \frac{1}{\rho_0}\big(\partial\rho/\partial t, \partial(\rho U)/\partial t, \partial(\rho V)/\partial t, \partial(\rho W)/\partial t, \partial(\rho E)/\partial t\big)^T \\
= {} & \frac{1}{\rho_0}\int \Big[\gamma_1 g_0 + \gamma_2\Big(\bar{a}^l u \mathrm{H}[u] + \bar{a}^r u(1 - \mathrm{H}[u]) + \bar{b}v + \bar{c}w\Big)g_0 \\
& + \gamma_3\Big(\mathrm{H}[u]g^l + (1 - \mathrm{H}[u])g^r\Big) \\
& + \gamma_4\Big((a^l u + b^l v + c^l w)\mathrm{H}[u]g^l + (a^r u + b^r v + c^r w)(1 - \mathrm{H}[u])g^r\Big) \\
& + \gamma_5\big((a^l u + b^l v + c^l w + A^l)\mathrm{H}[u]g^l \\
& + (a^r u + b^r v + c^r w + A^r)(1 - \mathrm{H}[u])g^r\big)\Big]\psi_\alpha d\mathbf{u},
\end{aligned}
\tag{2.28}
$$

for the determination of $\overline{A}$, where

$$\gamma_0 = \Delta t - \tau(1 - e^{-\Delta t/\tau}),$$
$$\gamma_1 = -(1 - e^{-\Delta t/\tau})/\gamma_0,$$
$$\gamma_2 = \left(-\Delta t + 2\tau(1 - e^{-\Delta t/\tau}) - \Delta t e^{-\Delta t/\tau}\right)/\gamma_0,$$
$$\gamma_3 = (1 - e^{-\Delta t/\tau})/\gamma_0,$$
$$\gamma_4 = \left(\Delta t e^{-\Delta t/\tau} - \tau(1 - e^{-\Delta t/\tau})\right)/\gamma_0,$$
$$\gamma_5 = -\tau(1 - e^{-\Delta t/\tau})/\gamma_0.$$

The above multidimensional GKS is similar to the directional splitting GKS (Xu 2001), except for additional terms $b^l, b^r, c^l, c^r, \overline{b}$, and $\overline{c}$ related to the flow variations in the tangential directions. In practice, instead of using Eq. (2.28), in many computations, $\overline{A}$ can be simply evaluated by

$$M_{\alpha\beta}^0 \overline{A}_\beta = -\frac{1}{\rho_0} \int [\overline{a}^l u \mathrm{H}[u] + \overline{a}^r u (1 - \mathrm{H}[u]) + \overline{b} v + \overline{c} w] \psi_\alpha g_0 d\mathbf{u},$$

which uses the kinetic-type smooth flow approximation from the spatial derivative to the time derivative for the local equilibrium state.

Finally, the time-dependent numerical fluxes at the point $(0,0,0)$ in the normal-direction across the cell interface can be computed by

$$
\begin{aligned}
\mathbf{F_W} &= \int_0^{\Delta t} \int u \psi f(0,0,0,t,u,v,w) S dt \, d\mathbf{u} \\
&= \int u \psi \Big[\Big(q_1 g_0 + q_2 \big(\overline{a}^l u \mathrm{H}[u] + \overline{a}^r u (1 - \mathrm{H}[u]) + \overline{b} v + \overline{c} w \big) g_0 + q_3 \overline{A} g_0 \Big) \\
&\quad + q_4 \Big(\mathrm{H}[u] g^l + (1 - \mathrm{H}[u]) g^r \Big) \\
&\quad + q_5 \Big((u a^l + v b^l + w c^l) \mathrm{H}[u] g^l + (u a^r + v b^r + w c^r)(1 - \mathrm{H}[u]) g^r \Big) \\
&\quad + q_6 \Big((u a^l + v b^l + w c^l + A^l) \mathrm{H}[u] g^l \\
&\quad + (u a^r + v b^r + w c^r + A^r)(1 - \mathrm{H}[u]) g^r \Big) \Big] S d\mathbf{u},
\end{aligned}
\tag{2.29}
$$

where $f(0,0,0,t,u,v,w)$ is given by Eq. (2.26) and the coefficients are

$$
\begin{aligned}
q_1 &= \Delta t - \tau(1 - e^{-\Delta t/\tau}), \\
q_2 &= 2\tau^2(1 - e^{-\Delta t/\tau}) - \tau \Delta t - \tau \Delta t e^{-\Delta t/\tau}, \\
q_3 &= \frac{\Delta t^2}{2} - \tau \Delta t + \tau^2(1 - e^{-\Delta t/\tau}), \\
q_4 &= \tau(1 - e^{-\Delta t/\tau}), \\
q_5 &= \tau \Delta t e^{-\Delta t/\tau} - \tau^2(1 - e^{-\Delta t/\tau}), \\
q_6 &= -\tau^2(1 - e^{-\Delta t/\tau}).
\end{aligned}
\tag{2.30}
$$

The evaluation of the moments of a Maxwellian distribution function can be found in the Appendix.

The above fluxes can be transformed into any mesh orientation in a finite volume scheme, even with moving and unstructured mesh (Jin & Xu 2007; Ren et al. 2016).

The modelling of the solution f at a cell interface is based on the integral solution of the BGK model, where a unit Prandtl number is recovered in the hydrodynamic limit. In order to simulate the flow with any Prandtl number, the heat transport part in the energy flux can be easily modified according to the correct Prandtl number. With a gas distribution function f at the cell interface in Eq. (2.26), the time-dependent heat flux can be evaluated precisely,

$$q = \frac{1}{2} \int \int_0^{\Delta t} (u - U)\left((u - U)^2 + (v - V)^2 + (w - W)^2\right) f \, S dt \, d\mathbf{u}, \qquad (2.31)$$

where U, V and W are averaged velocities. Then, the Prandtl number in GKS can be modified to any value by modifying the heat transport in the energy flux,

$$F_{\rho E}^{new} = F_{\rho E} + \left(\frac{1}{\Pr} - 1\right) q, \qquad (2.32)$$

where $F_{\rho E}$ is the energy flux in Eq. (2.29). Besides the above heat flux modification, in order to get the correct Prandtl number, the integral solution of Shakhov or ES-BGK models can be directly used as well. The Shakhov and ES-BGK models can be further combined to get a general unified model (Chen et al. 2015). To compute the continuum flow without updating the gas distribution function, the above GKS flux for the update of macroscopic flow variables is very accurate for the NS solution.

Once a cell size is able to resolve the NS solution, such as the laminar boundary layer, the initial reconstruction should present a continuous flow distribution across a cell interface. The collision time τ in the well-resolved case can be naturally defined by

$$\tau = \mu/p, \qquad (2.33)$$

where μ is the dynamical viscosity coefficient and p is the pressure. This is a well-known result of the Chapman-Enskog expansion of the kinetic BGK model (Vincenti & Kruger 1965). For the viscosity coefficient, μ can take any reasonable value in the determination of τ, such as Sutherland's law,

$$\mu = \mu_\infty \left(\frac{T}{T_\infty}\right)^{3/2} \frac{T_\infty + S}{T + S},$$

where T_∞ and S are temperatures with the values $T_\infty = 285k$ and $S = 110.4k$. The above particle collision time is implemented in GKS in the following way. The collision time in the initial distribution function f_0 is determined from the reconstructed macroscopic flow variables at the left- and right-hand sides of a cell interface. The collision time is based on $\mathbf{W}_0$ in Eq. (2.21), such as $\mu(\mathbf{W}_0)$ and $\tau = \mu(\mathbf{W}_0)/p$, where p is the pressure and is also a function of $\mathbf{W}_0$.

The physical shock thickness is related to the physical viscosity. The Navier-Stokes shock structure can be accurately obtained if the cell size is fine enough to resolve the shock structure (Xu 2001). However, the physical shock thickness is on the order of particle mean free path, which is in the kinetic length scale. For the continuum flow computation, the mesh size used is much larger than the particle mean free path. Therefore, the numerical cell size cannot fully resolve the shock structure. As a result, the physical shock structure must be replaced by a numerical one with the mesh size thickness. This is the common practice for any shock-capturing scheme. In such a situation, it is meaningless to talk about the solution of the original NS equations because there is no enough cell resolution to resolve such a solution. In order to get a numerical shock structure with a physically consistent mechanism, an enhanced viscosity coefficient is needed to enlarge the shock thickness to the mesh size. So, once there is a discontinuity in the initial reconstruction of macroscopic flow variables around a cell interface, it indicates that the cell resolution is not fine enough to resolve the flow structure and the structure has to be replaced somehow by a numerical one. In the un-resolved discontinuity region, an additional numerical dissipation is introduced in GKS. According to the magnitude of pressure jump at the cell interface, the collision time τ in GKS is defined as

$$\tau = \frac{\mu(\mathbf{W}_0)}{p(\mathbf{W}_0)} + \alpha_n \frac{|p_l - p_r|}{|p_l + p_r|} \Delta t, \tag{2.34}$$

where Δt is the CFL time step, which is related to the cell size by the CFL condition. The second term corresponds to the numerical viscosity. α_n is an adjustable parameter on the order of 1 to 5. The pressure jump at the cell interface is from the reconstructed macroscopic variables $\mathbf{W}^l$ and $\mathbf{W}^r$. In the smooth flow region or around the slip line, the additional collision time is very small or diminishes due to the continuous pressure distribution. Based on the numerical tests, the enhancement of the particle collision time does not deteriorate the boundary layer calculations once the layer is well resolved, but enhances the robustness of the GKS in its shock-capturing capability. The numerical dissipation in GKS through the inclusion of additional collision time enhances the physical mechanism of non-equilibrium particle free transport in f_0. This is consistent with physical reality within a shock layer, where the particle free transport effect is important for maintaining the non-equilibrium distribution function and the determination of a physical shock structure. For example, a particle may only take a few collisions in the passage from the upstream to the downstream. The particle free transport provides the kinetic type dissipation to stabilize the numerical shock layer. A non-equilibrium discontinuous gas distribution function f_0 with an enhanced free transport enlarges the shock thickness from the mean free path to the mesh size scale with the relation $\tau \sim \Delta t \sim \Delta x$ in the shock region. Therefore, even for the continuum flow computation, the GKS still uses the dynamics from the kinetic scale in the discontinuous shock region and avoids the shock instability (Ohwada et al. 2013). For example, the real gas distribution function used in the flux evaluation inside the numerical shock structure keeps a two-peaks structure, which represents the particles coming from upstream and downstream inside a numerical shock layer (Xu 2015). A significant amount of testing has been conducted using the GKS for the continuum flow computations. The GKS is especially suitable for the hypersonic flow simulation. As an enhanced GKS flow solver (Luo & Xu 2013), the collision time defined in Eq. (2.33) can be regarded as the physical one τ_p and the one in Eq. (2.34) can be defined as the numerical one τ_n. The coefficients $C_1, C_2, ..., C_6$ in Eq. (2.27) for the determination of the gas distribution function can keep τ_n in the exponential function part, $\exp(-\Delta t/\tau_n)$, and keep a physical one τ_p in other places.

2.3 Properties and Extensions of GKS

(a) Dynamic Differences among Various Kinetic Solvers

There are many kinetic solvers based on kinetic equations. The main difference between the GKS and other kinetic methods is the flux evaluation. Many kinetic schemes solve the collisionless Boltzmann equation for the flux evaluation, the so-called upwind approach according to the particle velocity, such as the schemes in (Pullin 1980; Deshpande 1986; Perthame 1992; Aristove 2012). Here we are going to analyse the dynamic differences in different flux construction. It helps to understand the difficulties for other kinetic solvers in obtaining accurate NS solutions, such as the calculation of the laminar boundary layer and their extensions to multi-scale methods.

In a smooth, well-resolved flow region, based on high-order reconstruction, the initial discontinuous jumps disappear, and the terms in the initial distribution function f_0 have $g^l = g^r$, $a^l = a^r$, $b^l = b^r$, and $c^l = c^r$. Consequently, Eq. (2.21) gives $g_0 = g^l = g^r$, and Eq. (2.22) and (2.23) reduce to $\bar{a}^l = \bar{a}^r = a^l = a^r$, $\bar{b}^l = \bar{b}^r = b^l = b^r$, and $\bar{c}^l = \bar{c}^r = c^l = c^r$. As a result, $\bar{A}$ is exactly equal to A^l and A^r in Eq. (2.28). Therefore, without any further assumption, the gas distribution function f at a cell interface in Eq. (2.26) automatically reduces to

$$f = [1 - \tau(u\bar{a} + v\bar{b} + w\bar{c} + \bar{A}) + t\bar{A}]g_0, \tag{2.35}$$

where $-\tau(u\bar{a} + v\bar{b} + w\bar{c} + \bar{A})g_0$ is the non-equilibrium state in the Chapman-Enskog expansion of the BGK model for the NS dissipative terms, and $g_0\bar{A}t$ is the time evolution part of the gas distribution function. Since the above time evolution part is obtained from the compatibility condition, $\int \psi(u\bar{a} + v\bar{b} + w\bar{c} + \bar{A})g_0 d\mathbf{u} = 0$, which is equivalent to the Euler equations, the GKS in the smooth region has a second-order accuracy in terms of space and time for the inviscid flow calculation and first-order accuracy in terms of time accuracy for the viscous terms. The above formulation is identical to the Lax-Wendroff-type central difference scheme for the Euler equations once we ignore the term $-\tau(u\bar{a} + v\bar{b} + w\bar{c} + \bar{A})g_0$ in Eq. (2.35). The solution in the smooth flow region shows that the GKS will get back to the central difference method once the initial discontinuity disappears. This is definitely a preferred property for any shock-capturing scheme, because in the smooth flow region, the central difference scheme is more accurate than the Riemann solver–based upwind scheme. The above solution for smooth flow can be even obtained from discontinuous initial data once $\Delta t \gg \tau$, because the contribution from f_0 decays exponentially with $\exp(-\Delta t/\tau)$ and the integration of the equilibrium state contributes mainly in the flux function, which gets to the distribution function of Eq. (2.35). This clearly indicates that the GKS will not be sensitive to the initial data reconstruction. Dynamically, the time evolution solution at the cell interface will quickly converge the physical solution in the continuum flow regime, even though the initial condition is not properly reconstructed. For example, in the laminar boundary layer computation at a high Reynolds number, even with the van Leer limiter for the initial data reconstruction, the GKS can give accurate NS solutions with only three or four grid points inside the boundary layer, which is difficult for many other Riemann solver-based shock-capturing schemes with the same initial data reconstruction.

For many other kinetic solvers, the upwind approach is directly used according to particle velocity in the flux construction at a cell interface. This is equivalent to solving the collisionless Boltzmann equation in the flux evaluation. One of the outstanding schemes in this category is the kinetic flux vector splitting (KFVS) scheme (Pullin 1980; Deshpande 1986). Based on the collisionless Boltzmann equation, i.e.,

$$f_t + uf_x + vf_y + wf_z = 0,$$

and the same initial data reconstruction of GKS, without particle collision the gas distribution function at a cell interface in Eq. (2.26) goes to

$$
\begin{aligned}
f &= f_0(-ut, -vt, -wt) \\
&= [1 - \tau(ua^l + vb^l + wc^l + A^l) - t(ua^l + vb^l + wc^l)]\mathrm{H}[u]g^l \\
&\quad + [1 - \tau(ua^r + vb^r + wb^r + A^r) - t(ua^r + vb^r + wc^r)](1 - \mathrm{H}[u])g^r \\
&= [1 - (\tau + t)(ua^l + vb^l + wc^l + A^l) + tA^l]\mathrm{H}[u]g^l \\
&\quad + [1 - (\tau + t)(ua^r + vb^r + wc^r + A^r) + tA^r](1 - \mathrm{H}[u])g^r.
\end{aligned}
\tag{2.36}
$$

This is the same scheme as the KFVS for the NS solution (KFVS-NS; Chou & Baganoff 1997). In the smooth flow region, the above equation is

$$f = [1 - (\tau + t)(u\bar{a} + v\bar{b} + w\bar{c} + \bar{A}) + t\bar{A}]\, g_0. \tag{2.37}$$

When compared with Eq. (2.35), it becomes apparent that in the smooth flow region, the above equation solves the NS equations with a dynamical viscosity coefficient $\mu_{kfvs-ns} = (\tau + t)p$ instead of $\mu_{ns} = \tau p$. As a result, additional numerical dissipation, which is proportional to t, or $\Delta t/2$ on the average, is introduced into KFVS-NS. The above free transport mechanism in flux evaluation is intrinsically rooted in many other kinetic solvers, including implicit-explicit (IMEX) Runge-Kutta (RK) and AP schemes (Filbet & Jin 2010), where the NS solution can be hardly obtained from this kind of scheme in the continuum flow regime.

When free transport is used in the flux evaluation, the additional numerical viscosity coefficient has the form of $\mu_{num} \simeq p\Delta t/2$, where p is the pressure and Δt is the time step. Since the time step is determined by the CFL condition, the artificial dissipation can be written as

$$\mu_{num} \sim (\rho/(\gamma(1 + M)))c\Delta x \sim \rho c\Delta x,$$

where Δx is the cell size, c is the sound speed, and M is the Mach number. The scheme with the above numerical dissipation can behave differently at the shock layer with the thickness on the particle mean free path $\sim \ell$ and the boundary layer with the thickness on the order $\sim \ell^{1/2}$. Let's use $N = 10$ cells to resolve a shock structure and a boundary layer. Since the shock thickness is proportional to the particle mean free path ℓ, in order to resolve the shock structure, the mesh size must satisfy $\Delta x \sim \ell/N$. Then, in the stationary shock case, due to the conditions $M \geq 1$ and $Re \approx 1$, the ratio between the numerical and physical viscosity coefficient is

$$\frac{\mu_{num}}{\mu_{phys}} \simeq \frac{c\Delta x Re}{\ell U} \simeq \frac{Re}{M}\left(\frac{\Delta x}{\ell}\right) \simeq \frac{Re}{M}\left(\frac{1}{N}\right) \ll 1.$$

Therefore, even with the free transport mechanism, these schemes can still produce an accurate NS shock layer once the structure is well resolved, because the contribution from the numerical dissipation is much less than the physical one. In the rarefied flow regime, in the shock structure calculation, many kinetic solvers, such as the DSMC and DVM, provide accurate solutions, even with the decoupled particle free transport and collision. This mainly comes from the small mesh size in comparison with the particle mean free path. In other words, when the cell size is less than the particle mean free path, which is used to resolve the shock structure, the particle takes mostly collisionless movement on its path passing through the cell interface. This is also one of the main reasons why many kinetic solvers need the cell size and time step to be less than the particle mean free path and collision time. Providing an accurate NS shock structure does not mean that the kinetic solver is valid for the viscous flow computation because there are different kinds of flow structures there. These kinds of kinetic solvers with collisionless flux function can behave differently for the boundary layer computation in the continuum flow regime. The laminar boundary layer thickness is proportional to $\ell^{1/2}$, which can be much larger than the particle mean free path in cases with high Reynolds numbers. When the boundary layer thickness l_b is resolved with the same number of grid points, such as $l_b = N\Delta y$, there is

$$\frac{\mu_{num}}{\mu_{phys}} \simeq \frac{c\Delta y Re}{LU} \simeq \frac{Re}{M}\left(\frac{\Delta y}{L}\right) \simeq \frac{\sqrt{Re}}{M}\left(\frac{1}{N}\right),$$

where L is the length of the flat plate. The boundary layer thickness $l_b \simeq \sqrt{\nu L/U}$ has been used in the above approximation. Therefore, for a subsonic boundary layer with $M \ll 1$ and $Re \gg 1$, i.e., $M = 0.3$ and $Re = 10^5$, the numerical dissipation could dominate the physical one with $\mu_{num} >> \mu_{phys}$ when the boundary layer is resolved with $N = 10$ grid points. It is not surprising to notice that many kinetic solvers need 60 to 100 grid points to give a reasonable boundary layer solution, which is the same as the Flux Vector Splitting (FVS) scheme, in order to suppress the time step–related numerical dissipation.

The ratio between boundary layer thickness l_b and the particle mean free path ℓ is

$$\frac{l_b}{\ell} = \frac{\sqrt{Re}}{M}.$$

So, for a laminar boundary layer at $Re = 10^5$ and $M = 0.3$, the boundary layer thickness can easily go to a 100 particle mean free path. Most NS solvers starting from macroscopic governing equations directly can get an accurate boundary layer solution with 10 grid points, and each cell size can have tens or hundreds of particle mean free paths. The GKS can get the same performance as the standard NS solver and give accurate solutions with a few grid points within the layer. The multi-scale method presented in later sections can give an accurate NS solution as well, which is shown in the boundary layer test in Section 4. In the continuum flow regime, due to particle collision, the contribution from the initial gas distribution function f_0 decays exponentially and the equilibrium state will make the dominant contribution in the final flux function. Unfortunately, many kinetic solvers keep the free transport of f_0 in the flux evaluation, even when the cell size is much larger than the particle mean free path. In such a situation, the free transport is definitely inconsistent with the physical reality, and the particle would not take free movement across a length scale with tens or hundreds of particle mean free paths. The mistake in these kinetic solvers comes from the direct application of free streaming transport of the Boltzmann equation without recognizing the difference between the kinetic scale modelling in the Boltzmann equation and the numerical mesh size and time step scales.

In the lattice Boltzmann method (LBM; Chen & Doolen 1998), due to the use of a regular lattice and in a low-speed incompressible flow simulation, the numerical dissipative coefficient from the free transport has a fixed value and is luckily absorbed into the physical one. For example, the real viscosity coefficient in LBM formulation is proportional to $\tau_{equivalent} = \tau + \Delta t/2$, as presented in Eq. (2.37) due to the free transport mechanism. In order to simulate flow with 'physical' viscosity $\tau_{equivalent}$, the collision time used in LBM has to be defined by

$$\tau = \tau_{equivalent} - \Delta t/2.$$

This lucky coincidence is purely from the regular lattice and fixed numerical dissipative coefficient of the incompressible flow limit with a constant temperature. If a non-uniform mesh is used, the LBM will encounter great difficulty in controlling its numerical dissipation. Even for the laminar boundary layer, we can hardly find any LBM simulation due to the singular leading edge of the boundary layer solution, where a uniform mesh can hardly be applied in this case. The coupled transport and collision in the GKS flux of Eq. (2.35) removes the additional numerical dissipation from the particle free transport, and the laminar boundary layer solution can be obtained accurately even with a few grid points (Xu 2001). Recently, with the same discretization of particle velocity space as LBM, a discrete unified gas-kinetic scheme (DUGKS) has been developed by including particle collision in the flux evaluation (Guo et al. 2013; Guo et al. 2015). The DUGKS does not have the numerical dissipative term ($\sim \Delta t/2$) of LBM and presents accurate NS solutions even with unstructured mesh (Zhu et al. 2016). In other words, the DUGKS promotes the LBM to a new level of accuracy and robustness, which solves many of the problems associated with LBM (Succi 2015), such as the use of uniform mesh, slow convergence for steady flows, limitations of the Mach number, and difficulties in actual coding for the implementation of complex boundary conditions.

(b) Comparison of a GKS Flux Function and Exact Riemann Solver

When starting from discontinuous initial data, most CFD methods for compressible flow simulation are based on the Riemann solver for the flux evaluation, i.e., the so-called Godunov method. In order to understand the differences between the GKS and Godunov methods, it is helpful to compare their flux functions and underlying mechanisms. The GKS method has an outstanding record in terms of shock capturing. Even in the case of continuum flow computation, the inclusion of non-equilibrium transport seems necessary in the construction of a reliable shock-capturing scheme.

Based on the discontinuous initial data, as shown in Figure 2, the GKS and Riemann solvers take on different modelling in the construction of an evolution solution at a cell interface. The GKS covers a gas evolution process from the kinetic to the hydrodynamic scales. It first presents particle free transport at $t \le \tau$, then with the account of particle collision, the initial non-equilibrium distribution function approaches an equilibrium at $t \ge \tau$. With the intensive particle collisions $t \gg \tau$, the quasi-equilibrium NS distribution function emerges from the accumulating evolution (integration) of the equilibrium state, providing the NS flux function. The Euler solution becomes a limiting solution with the absence of viscous terms. The dynamics used in the GKS flux function depends on the cell Knudsen number ($Kn_C = \tau/\Delta t$). In the smooth region at $Kn_C \ll 1$, the GKS becomes a standard NS flow solver. However, in the discontinuous region, such as in a numerical shock layer, the GKS incorporates the physics of $Kn_C \sim 1$ and uses the kinetic scale particle free transport for providing numerical

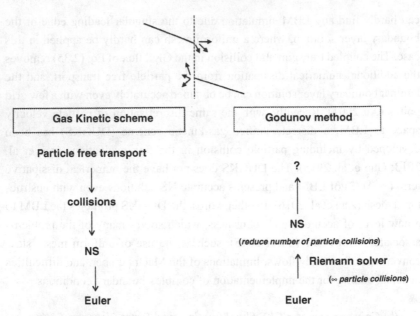

Figure 2 Physical process from a discontinuity in GKS and Godunov-type method

dissipation in the construction of a numerical shock structure with a mesh size thickness. The introduction of an additional term in the definition of particle collision time in Eq. (2.34) serves to enlarge the free transport mechanism in GKS for capturing the numerical shock structure.

On the other hand, for the Godunov method, the Riemann solver provides an exact solution to the Euler equations. The infinite number of particle collision is assumed to take place instantaneously at $t = 0$ in order to generate distinctive wave structures, such as shock, contact, and rarefaction waves. For example, the existence of a contact wave in the Riemann solution requires intensive particle collision to prevent particles from penetrating the contact or slip line. Under the numerical PDE methodology, it is fully legitimate to use the Riemann solver in the flux evaluation for the Euler solutions, because the Riemann solver gives an exact solution of the equations under two constant initial states. However, the Euler equations are idealized equations constructed in the continuous space and time with the equilibrium state assumption. The shock wave of the Euler equations has a zero thickness, but the numerical computation takes place in a discretized space and time with limited cell size. The CFD should target the modelling and simulation of physical laws in a discretized space with limited cell resolution. The physical effect from the limited mesh size and time step needs to be incorporated into the design of a numerical scheme. In reality, the flow is composed of particles, and there is a physical viscosity related to the particle collision time $\mu = \tau p$. Starting from a discontinuity, at time $t = 0$, equilibrium states in the Riemann solution cannot be immediately achieved via initially non-equilibrium discontinuous data, especially inside a numerical shock layer with a highly non-equilibrium state. Furthermore, following the methodology of the Godunov method, if the NS solution is required, the physical dissipation has to be added after the Riemann solution. This is equivalent to reducing the number of particle collisions in order to keep a slightly non-equilibrium state. In terms of a dynamic system, this evolution process from equilibrium to non-equilibrium would not exist. This is contradictory to the second law of thermodynamics in the gas evolution process around a cell interface. Thus the dynamic process in the Godunov method for the NS solution has intrinsic physical inconsistency. Practically, the conventional CFD method uses an operator splitting approach to treat the inviscid and viscous terms separately, which basically requires different initial states to take into account inviscid wave propagation and viscous diffusion.

In order to capture the shock transition in the mesh size scale, the Godunov method needs additional numerical dissipation, but the Riemann solution does not provide such a numerical dissipative mechanism. For the Godunov method, the numerical dissipation solely arises from the initial data reconstruction, such that the kinetic energy is converted into thermal energy in the averaging process through the introduction of initial discontinuities (Xu & Li 2001). This kind of numerical dissipation depends closely on the relative orientation between the mesh and shock layer. The carbuncle phenomenon seems unavoidable in the Godunov method, since it is impossible to expect a purely artificially mesh-related dissipation to provide an appropriate physical mechanism in constructing a stable numerical shock structure all the time. For the GKS, the numerical dissipation in the shock region is added explicitly through an enhanced particle collision time for enlarging particle free transport to mimic the physical mechanism in the non-equilibrium shock layer.

The Riemann solver–based methods are difficult to use for the development of compact high-order schemes (order ≥ 3) for the flow simulation with strong shock interaction because the high-order schemes should have a minimal amount of numerical dissipation in the initial reconstruction. The implicit and uncontrollable artificial dissipation in the Godunov method may not be reliable in the flow simulation with shocks. At the same time, with increasing computational power, the mesh size used can become very small. For the hypersonic viscous flow computation, the mesh size used next to the surface may get to the order of the particle mean free path. With the use of such a small mesh size, there is no infinite number of collisions to form the so-called Riemann solution within a time step. In fact, with the large variation of the cell size and the local cell Knudsen number, the CFD is intrinsically a multi-scale problem for flow modelling and computation. The traditional numerical PDE approach targeting the numerical discretization of PDE can hardly be extended to develop a multi-scale method, as there is no multi-scale PDE for gas dynamics.

(c) Multidimensional GKS Flux Function

In GKS, on both sides of a cell interface, the initially reconstructed flow variables keep derivatives in both normal and tangential directions. The flow variations in all directions will participate in the gas evolution and determine the time-dependent solution at a cell interface, such as the contributions from $ua^l\,g^l$ and $ua^r\,g^r$ in the normal direction, and $vb^l\,g^l$, $vb^r\,g^r$, $wc^l\,g^l$, $wc^r\,g^r$ in the tangential directions in Eq. (2.26). Therefore, even with a fixed normal direction of the cell interface, the particle transport (from f_0) and wave propagation (from the integration of the equilibrium state g) can go in any direction relative to the mesh orientation. Therefore, the GKS is intrinsically a multidimensional scheme, and it will not be sensitive to the quality of the mesh if the initial reconstruction can be properly obtained. On the contrary, the Riemann solution is an exact solution of the Euler equations with two constant states. The waves from the initial discontinuity propagate in the normal direction of a cell interface only. Therefore, the local wave propagating direction in the Godunov method is solely determined by the mesh orientation. It will not be surprising to see the difficulties for the Godunov method in capturing the laminar boundary layer solution with unstructured mesh everywhere and the sensitivity of the post-shock oscillation relative to the mesh orientation in the multidimensional case. However, the GKS can use unstructured mesh everywhere, even inside the boundary layer, and get accurate NS solutions (Ji et al. 2020).

(d) Smooth Transition from Upwind to Central Difference

The GKS has a multi-scale evolution process in the flux construction, and the flux makes a smooth transition from the collisionless particle transport in f_0 to the central difference discretization in the integration of the equilibrium state g. As a result, with the proper control of particle collision time in Eq. (2.34), the kinetic scale upwind mechanism is used to capture the numerical shock wave, and the hydro-dynamical central difference is used to recover a smooth NS solution. So, the GKS will not be sensitive to the initial reconstruction in the viscous flow computation, especially for high Reynolds flow computation, because the scheme will automatically and quickly make a smooth transition from the initial discontinuity to a nearly smooth NS flow solver when $\Delta t \gg \tau$. In a barely resolved boundary layer, even a constant state inside each cell without any initial reconstruction, such as first-order scheme reconstruction, the NS solution can be still obtained reasonably due to the automatic diminishing of the contribution from f_0 and the emerging NS distribution from the integration of g. As pointed out in Ohwada & Kobayashi (2004), the transition from f_0 to g is the key to the success of the GKS in capturing both discontinuous and continuous solutions. The kinetic scale particle free transport has a physical mechanism to remove the shock instability in the shock-capturing scheme (Ohwada et al. 2013). In practice, the flux vector splitting (FVS) upwind and the Lax-Wendroff-type central difference scheme have been unified under the GKS framework. The advantages of the upwind scheme for shock capturing and the central difference for smooth viscous solutions are maintained in the GKS in different regions. Most traditional CFD methods are based on either upwind or central difference in algorithm development.

(e) High-Order Extension and Compact GKS with Spectral Resolution

In the development of high-order GKS, the same integral solution (2.11) is used for the flux evaluation, but the initial distribution f_0 and the equilibrium state g are determined from high-order reconstructed macroscopic flow variables. Based on high-order initial data reconstruction, the corresponding high-order Chapman-Enskog expansion for the NS gas distribution function can be obtained. With a third-order expansion of f_0 and g, a time-dependent gas evolution solution has been obtained for the flux evaluation. For example, a directional splitting third-order GKS flux was constructed for the compressible flow computation (Li et al. 2010), followed by a multidimensional one (Luo & Xu 2013). Equipped with weighted essentially non-oscillatory (WENO) reconstruction, the WENO-type GKS has been developed and compared with traditional Riemann solver-based WENO methods for the inviscid and viscous flow computations (Kumar et al. 2013; Luo et al. 2013; Ji et al. 2018a; Ji et al. 2018b).

Recently, a class of compact high-order GKS with spectral-like resolution has been developed on structured and unstructured meshes (Zhao et al. 2019; Zhao et al. 2020a; Zhao et al. 2020b). In GKS, the time-dependent gas distribution function at a cell interface can provide not only the flux function and its time derivative, but also the time accurate solution $f(t^{n+1})$ on the cell interface at the beginning of the next time step, from which the macroscopic flow variables can be evaluated. As a result, besides updating the conservative flow variables inside each control volume through the interface fluxes, the cell averaged gradients of flow variables $\nabla \mathbf{W}$ inside the cell can be updated by applying the divergence theorem, such as integrating the flow variables around the closed cell interfaces of a control volume Ω,

$$\oiint \mathbf{W} d\vec{S} = \nabla \mathbf{W} \Omega.$$

As a result, inside each control volume, both the cell-averaged flow variables W^{n+1} and their cell-averaged gradients ∇W^{n+1} can be updated. Here, the flow variables and their gradients are coming from the same physical evolution solution at the cell interfaces. Based on both cell averaged flow variables and their gradients, compact sixth-order and eighth-order linear and nonlinear initial reconstructions have been developed (Zhao et al. 2019). Both schemes achieve a better spectral-like resolution in cases of large wave numbers compared with those from other well-established compact schemes with globally coupled flow variables and derivatives, such as the compact schemes of Lele (Lele 1992). In GKS, in order to avoid spurious oscillation in the discontinuous region, the compact high-order linear reconstruction from the symmetric stencil can be divided into sub-stencils and apply a biased nonlinear WENO-Z reconstruction. The initial non-equilibrium state f_0 is obtained based on the nonlinear WENO-Z reconstruction for macroscopic flow variables, and the equilibrium state g is constructed from the linear symmetric reconstruction. This property gives GKS a significant advantage for capturing both discontinuous shock waves and linear aero-acoustic waves in a single computation through its dynamic adaptation of nonlinear and linear reconstructions for the non-equilibrium and equilibrium states. This adaptive model helps to solve a long-held problem in the development of high-order CFD methods about the choices of the linear and nonlinear reconstructions. The high-order compact GKS unifies these two reconstructions in a single numerical scheme through a relaxation process rather than any predefined numerical hybrid approach. In the unstructured mesh case, the high-order reconstructions for f_0 and g are based on the least-square and WENO-Z methods (Zhao et al. 2020b).

In high-order algorithms development for CFD, the compact GKS uses the compact stencil, achieves sixth-order and eighth-order accuracy, uses a reasonable time step with CFL number ≥ 0.3, has almost the same robustness as a second-order scheme, and obtains accurate solutions in both shock and smooth regions without introducing any troubled cell detection and additional limiting processes. The nonlinear reconstruction in the compact GKS is based on the well-studied WENO-Z technique. At the same time, the compact GKS solves the Navier-Stokes equations automatically due to its combined inviscid and viscous terms in the time-dependent gas distribution function. This property is preferred in the flow simulation using unstructured mesh, where there is no need to get separate reconstructions for the inviscid and viscous terms. Due to the use of a two-stage fourth-order time-stepping technique in the solution update (Li & Du 2016; Pan et al. 2016), in order to achieve a fourth-order time accuracy, the GKS needs only two stages instead of four in the traditional Runge-Kutta method. As a result, the GKS saves two reconstructions in comparison with the other Runge-Kutta-based high-order schemes. The high-order GKS presents state-of-art numerical solutions for a wide range of flow problems, e.g., strong shock discontinuity, shear instability, aero-acoustic wave propagation, and NS solutions. It shows a new level of maturity in the development of high-order CFD methods. The success of GKS crucially depends on its high-order gas evolution model at a cell interface and its kinetic-hydrodynamic flow physics. The use of the first-order Riemann solver seems difficult for developing a compact numerical algorithm with all the properties of accuracy, efficiency, compactness, and robustness, especially for the flow simulations with a discontinuous shock wave.

Here an example is presented from the eighth-order compact GKS for the Euler solution. The test problem was extensively used for inviscid flow computation (Woodward & Colella 1984). The computational domain is $[0, 3] \times [0, 1]$. The height of the wind tunnel is 1, and the length is 3. The step is located at $x = 0.6$, with a height of 0.2 in the tunnel. The gas in the tunnel is a uniform flow with the initial condition $\rho = 1, U = 3, V = 0, p = 1/1.4$. In the current computation, there is no special treatment around the step corner. With the cell size $\Delta x = \Delta y = 1/240$, the density and vorticity distributions at output time $t = 4.0$ are shown in Figure 3, where accurate solutions have been obtained by the eighth-order compact GKS with a CFL number 0.5.

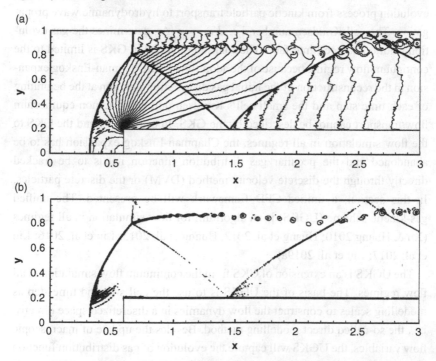

Figure 3 Step problem at Mach 3: density (a), vorticity (b) distributions at $t = 4.0$ by the eighth-order compact GKS on uniform mesh with $\Delta x = \Delta y = 1/240$. Courtesy of F.X. Zhao.

2.4 Summary

In this section, we present the basic kinetic theory and the gas-kinetic scheme for the Euler and Navier-Stokes solutions. In the continuum flow regime, the NS gas distribution function is well-defined through the Chapman-Enskog expansion. Similar to the other macroscopic flow solvers, only the macroscopic flow variables are updated in GKS through numerical fluxes, which are obtained from the time-dependent gas distribution functions at cell interfaces. The comparison between GKS and Riemann solver–based CFD methods is introduced. The GKS is a flow solver in the continuum flow regime, which becomes the limiting flow solver of the multi-scale methods introduced in the later sections.

3 The Unified Gas-Kinetic Scheme

3.1 Preface

In the previous section, we presented the finite volume gas-kinetic scheme (GKS) for the Navier-Stokes (NS) solution. Even for the continuum flow computation, the interface flux function is still based on a multi-scale gas evolution process from kinetic particle transport to hydrodynamic wave propagation. The cell Knudsen number, i.e., $Kn_C = \tau/\Delta t$, determines the gas evolution solution at a cell interface. The solution obtained in GKS is limited to the continuum flow regime because of the adoption of the Chapman-Enskog expansion in the reconstruction of the initial gas distribution function at the beginning of each time step and the small cell's Knudsen number. The non-equilibrium flow transport cannot be described by the GKS. In order to extend the GKS to the flow simulation in all regimes, the Chapman-Enskog expansion has to be abandoned and the peculiar gas distribution function needs to be tracked directly through the discrete velocity method (DVM) or the discrete particles. In this section, a unified CFD framework will be presented. The unified gas-kinetic scheme (UGKS) is an algorithm for flow simulation in all regimes (Xu & Huang 2010; Huang et al. 2012; Huang et al. 2013; Liu et al. 2016; Liu et al. 2017; Liu et al. 2019a).

The UGKS is an extension of GKS from the continuum flow simulation to all flow regimes. The basis of the UGKS is to use the cell size and time step as modelling scales to construct the flow dynamics in a discretized space directly, i.e., the so-called direct modelling method. Besides the update of macroscopic flow variables, the UGKS will capture the evolution of gas distribution function in a discretized particle velocity space with DVM formulation. In terms of physical modelling, the UGKS becomes simpler than the GKS because the kinetic theory, such as the Chapman-Enskog expansion for the NS solution, is not needed in the construction of the UGKS. The update of the gas distribution function makes the scheme more reliable in terms of the description of non-equilibrium flow. Unlike many other kinetic solvers for rarefied flow, such as the DSMC and direct Boltzmann solver, the cell size and time step used in the UGKS are not limited by particle mean free path and particle collision time. Multiple particle collision within a time step will be quantitatively evaluated in the scheme. The NS solution will be recovered automatically by the UGKS in the continuum flow regime with a small cell's Knudsen number $Kn_C \ll 1$. The gas dynamic equations of the UGKS depend on the cell's Knudsen number, or the modelling scale explicitly.

3.2 Discretization of the Particle Distribution Function

In DVM, in order to update the gas distribution function, the particle velocity space is discretized with grid points. The discretized physical space is divided into control volume, i.e., $\Omega_{i,j,k}(\mathbf{x}) = \Delta x \Delta y \Delta z$ with the cell sizes $\Delta x = x_{i+1/2,j,k} - x_{i-1/2,j,k}$, $\Delta y = y_{i,j+1/2,k} - y_{i,j-1/2,k}$, and $\Delta z = z_{i,j,k+1/2} - z_{i,j,k-1/2}$. The temporal discretization is denoted by t^n for the n_{th} time step. The particle velocity space is discretized by Cartesian mesh points with velocity spacing Δu, Δv, and Δw for each control volume $\Omega_{l,m,n}(\mathbf{u}) = \Delta u \Delta v \Delta w$ around the centre of (u_l, v_m, w_n) velocity. The flow variable is a volume-averaged gas distribution function in $\Omega_{i,j,k}\Omega_{l,m,n}$ at time step t^n,

$$
\begin{aligned}
f(x_i, y_j, z_k, t^n, u_l, v_m, w_n) &= f^n_{i,j,k,l,m,n} \\
&= \frac{1}{\Omega_{i,j,k}(\mathbf{x})\Omega_{l,m,n}(\mathbf{u})} \int_{\Omega_{i,j,k}} \int_{\Omega_{l,m,n}} f(x,y,z,t^n,u,v,w)d\mathbf{x}\,d\mathbf{u}.
\end{aligned} \tag{3.1}
$$

The selection of the discrete points and the range of the velocity space in DVM depend on the specific problem. In a finite particle velocity domain $\Omega(\mathbf{u}) \in [\mathbf{u}_{min}, \mathbf{u}_{max}]$, a set of points $\mathbf{u}_\alpha = (u_l, v_m, w_n)$ are used to discretize the space, where $l = 1, ..., n_x$, $m = 1, ..., n_y$, and $n = 1, ..., n_z$ are an index representing the velocity points (l, m, n). Once the value of f at a discrete velocity point $f_{i,j,k,l,m,n}$ is given, the macroscopic flow variables in the physical space can be obtained by the numerical quadrature integration in the particle velocity space. For the continuum and low transitional flow, the distribution function is close to a Maxwellian. Gaussian quadrature is used due to its accuracy and efficiency. For the strong non-equilibrium flow, Newton-Cotes integration is usually adopted with increased velocity points. The details of the distribution function with regard to discretized particle velocity are provided in the Appendix.

3.3 Unified Gas-Kinetic Scheme

The UGKS updates the gas distribution function directly instead of reconstructing it by applying the Chapman-Enskog expansion in GKS. In the UGKS, in order to model the gas evolution process, the gas-kinetic BGK model, the Shakhov model, the ES-BGK model, the Rykov model for diatomic gases, and even the full Boltzmann equation can be used in the construction of a multi-scale algorithm. Basically, a local time evolution solution of the gas distribution function at a cell interface and the particle collision inside each cell are modelled in the scales of cell size and time step. The kinetic model equation used in the construction of the UGKS for monatomic gas can be the Shakhov model (Shakhov 1968), which is

$$f_t + \mathbf{u} \cdot \nabla_{\mathbf{x}} f = \frac{f^+ - f}{\tau}, \tag{3.2}$$

and

$$f^+ = g[1 + (1 - \text{Pr})\mathbf{c} \cdot \mathbf{q}\left(\frac{\mathbf{c}^2}{RT} - 5\right)/(5pRT)] \tag{3.3}$$
$$= g + g^+,$$

where g is the Maxwellian, $\mathbf{q}$ is the heat flux, and Pr is the Prandtl number. In the Shakhov model, in order to fix the unit Prandtl number in the BGK model, a heat-flux based correcting term was added in the equilibrium state. Although the above kinetic model equation is much simpler than the full Boltzmann equation, it shares a similar asymptotic property in the hydrodynamic regime (Vincenti & Kruger 1965). Both equations recover the Euler and Navier-Stokes equations when the Knudsen number is small. The kinetic model equation approximates the full Boltzmann equation accurately in many flow cases. The full Boltzmann equation is a modelling equation in the particle mean free path and collision time scale. In such a kinetic scale, there is a certain difference between the kinetic model and the full Boltzmann collision term. However, the difference is not as large as one might think of (Xu & Liu 2017). For the unified scheme, the ratio of the time step Δt over the local particle collision time τ varies significantly from the kinetic scale regime $\Delta t \leq \tau$ to the hydrodynamic scale regime $\Delta t \gg \tau$. In the regime with $\Delta t \geq \tau$, the difference between the solution of the full Boltzmann equation and the kinetic model equation gradually diminishes (Liu et al. 2016). When a particle encounters multiple collisions within a time step, the evolution of a gas distribution function will not be sensitive to individual binary particle collision. Therefore, the use of the kinetic model equation is accurate enough in most applications. The UGKS with the implementation of the hybrid particle collision terms of the full Boltzmann equation and model equation has been constructed as well, where the full Boltzmann collision term is used in the local region when $\Delta t \leq \tau$ and the kinetic model equation is used in other regions with $\Delta t \geq \tau$. In the following, the UGKS for monatomic gas is based on the kinetic model equation. For diatomic gas with rotational and vibrational degrees of freedom, the corresponding UGKS is given in (Liu et al. 2014; Wang et al. 2017).

In a discretized phase space, the UGKS evolves the gas distribution function and the conservative flow variables in a finite volume Ω_x of physical space through the interface fluxes and the inner particle collision. The distribution function is updated by

$$f^{n+1} = f^n - \frac{1}{\Omega_x} \int_{t^n}^{t^{n+1}} \int_{\partial\Omega} \mathbf{u} \cdot \mathbf{n} f \, ds dt + \frac{1}{\Omega_x} \int_{t^n}^{t^{n+1}} \int_{\Omega_x} \frac{f^+ - f}{\tau} d\mathbf{x} dt, \qquad (3.4)$$

at each discretized particle velocity, and

$$\mathbf{W}^{n+1} = \mathbf{W}^n - \frac{1}{\Omega_x} \int_{t^n}^{t^{n+1}} \int_{\partial\Omega} \psi \mathbf{u} \cdot \mathbf{n} f \, d\mathbf{u} ds dt, \qquad (3.5)$$

for the update of macroscopic conservative flow variables. Both the microscopic gas distribution function and macroscopic conservative flow variables are evolved from t^n to t^{n+1}. In order to close the above equations, at the centre of a cell interface $\mathbf{x}_0$, the time-dependent gas distribution function $f(\mathbf{x}_0, t, \mathbf{u})$ and the time-dependent inner cell particle collision have to be modelled. Both the flux for the gas distribution function in Eq. (3.4) and the flux for the macroscopic flow variables in Eq. (3.5) depend on the same cell interface gas distribution function $f(\mathbf{x}_0, t, \mathbf{u})$ at each particle velocity point $\mathbf{u} = (u_l, v_m, w_n)$. Since the ratio of $\Delta t / \tau$ may change significantly at different regions in a computational domain, the evolution of the gas distribution function at the cell interface needs to capture different flow physics, from the kinetic scale evolution $\Delta t \leq \tau$ to the hydrodynamic one $\Delta t \gg \tau$. Purely free transport or the direct upwind implementation in the flux evaluation at a cell interface according to particle velocity will definitely fail in the development of a multi-scale method and it will be impossible to recover the NS gas distribution function for the numerical flux in the continuum flow regime at a small cell's Knudsen number. The numerical procedures in the UGKS are as follows. The first step is to reconstruct the gas distribution function at each discretized particle velocity and the macroscopic flow variables inside each cell in the physical space. Then, the time-dependent gas distribution function at the cell interface, which is the same as that in the GKS, is constructed from the integral solution of the kinetic model equation. Then, the time-dependent gas distribution function is used for the flux evaluation for the updates of both the gas distribution function and the conservative flow variables. After updating the macroscopic flow variables, the equilibrium state can be determined inside each control volume. Then, the non-equilibrium state is updated inside each cell through the interface fluxes and the explicit-implicit treatment of inner cell particle collision. The use of a time-dependent interface gas distribution function is the key for designing the multi-scale UGKS, where the interface solution provides the flow physics from the particle-free transport to the hydrodynamic wave propagation.

Depending on the scales of Δx and Δt, the solution at cell interface $f(\mathbf{x}_0, t, \mathbf{u})$ is modelled from an evolution solution of the kinetic model equation (3.2). Assume the centre of the cell interface is located at $\mathbf{x}_0 = 0$, and the beginning of each time step is $t^n = 0$, along the characteristic line, the integral solution at $\mathbf{x}_0 = 0$ is

$$f(0, t, \mathbf{u}) = \frac{1}{\tau} \int_0^t f^+(\mathbf{x}', t', \mathbf{u}) e^{-(t-t')/\tau} dt' + e^{-t/\tau} f_0(-\mathbf{u}t, \mathbf{u}), \tag{3.6}$$

where $\mathbf{x}' = -\mathbf{u}(t - t')$ is the particle trajectory and f_0 is the initial distribution function. The same evolution equation has been used in the construction of the interface solution in the GKS, such as Eq. (2.11) in the previous section. However, in the GKS, the initial distribution function f_0 is reconstructed based on the Chapman-Enskog expansion from the macroscopic flow variables. In the UGKS, the above f_0 is directly updated from the previous time step. The velocity distribution at each discretized particle velocity point is assumed to be linearly distributed inside each control volume and can be constructed as

$$f_0(\mathbf{x}, \mathbf{u}) = (f_0^l + \sigma^l \cdot \mathbf{x})(1 - H[\mathbf{x}]) + (f_0^r + \sigma^r \cdot \mathbf{x})H[\mathbf{x}], \tag{3.7}$$

where (f_0^l, f_0^r) are the initial distributions at the left- and right-hand sides of the cell interface at a particle velocity, and (σ^l, σ^r) are the reconstructed gradients. In the GKS, the corresponding f_0 is given in Eq. (2.13) for the NS solution.

Similar to Eq. (2.15) in the GKS, the local post-collision distribution function f^+ is expanded in space and time on both sides of a cell interface,

$$f^+(\mathbf{x}, t, \mathbf{u}) = g_0[1 + (1 - H[x])\bar{a}^l x + H[x]\bar{a}^r x + \bar{b}y + \bar{c}z + \bar{A}t] + g_0^+, \tag{3.8}$$

where g_0 and g_0^+ are the distribution at $t^n = 0$.

The functions g_0, g_0^+, and τ can be fully determined from macroscopic variables at the cell interface and time $t^n = 0$. The same reconstruction scheme of the GKS can be used to reconstruct $\mathbf{W}$ inside each control volume. The macroscopic flow variables on the cell interface at $t = 0$ are obtained by taking into account colliding particles from the initial distribution f_0, which is the same as Eq. (2.21) of the GKS, but the moments of the distribution function become the summation of discretized particle velocity points,

$$\mathbf{W}_0 = \sum \omega \psi \left(f_0^l \mathrm{H}[u] + f_0^r (1 - \mathrm{H}[u]) \right), \tag{3.9}$$

where ω is the weighting function of the numerical quadrature. The coefficients $\overline{a}^l, \overline{a}^r, \overline{b}, \overline{c}$, and $\overline{A}$ can be obtained from the slopes of macroscopic flow variables, which are identical to Eqs. (2.22)–(2.25) of the GKS. The time derivative of $\mathbf{W}$ in the determination of $\overline{A}$ in Eq. (3.8) can be calculated from the compatibility condition,

$$\int \psi g_t d\mathbf{u} = \int \overline{A} \psi g_0 d\mathbf{u} = \frac{\partial \mathbf{W}}{\partial t}$$
$$= -\int [(\overline{a}^l \mathrm{H}[u] + \overline{a}^r (1 - \mathrm{H}[u]))u + \overline{b}v + \overline{c}w] g_0 \psi d\mathbf{u}, \tag{3.10}$$

which is a simplification of Eq. (2.28) in the GKS. The moments of the equilibrium distribution function and the determination of all these coefficients can be found in the Appendix. Substituting Eq. (3.7) and Eq. (3.8) into the integral solution Eq. (3.6), the gas distribution function at the cell interface for each point in the particle velocity space is

$$\begin{aligned}
f(\mathbf{x}_0, t, \mathbf{u}) &= (1 - e^{-t/\tau})(g_0 + g_0^+) \\
&\quad + ((t + \tau)e^{-t/\tau} - \tau)\left(\overline{a}^r (1 - \mathrm{H}[u]) + \overline{a}^l \mathrm{H}[u] \right) u g_0 \\
&\quad + ((t + \tau)e^{-t/\tau} - \tau)(\overline{b}v + \overline{c}w) g_0 \\
&\quad + (t + \tau(e^{-t/\tau} - 1))\overline{A} g_0 \\
&\quad + e^{-t/\tau}(f_0^l - t\mathbf{u} \cdot \sigma^l)\mathrm{H}[u] \\
&\quad + e^{-t/\tau}(f_0^r - t\mathbf{u} \cdot \sigma^r)(1 - \mathrm{H}[u]) \\
&= [C_1 \widetilde{g}_0 + C_2 \widetilde{g}_{\overline{x}} \cdot \mathbf{u} + C_3 \widetilde{g}_t] + [C_4 \widetilde{f}_0 + C_5 \widetilde{f}_{0\overline{x}} \cdot \mathbf{u}] \\
&= f^{eq}(t) + f^{fr}(t), \tag{3.11}
\end{aligned}$$

where all coefficients $C_1, C_2, \ldots, C_5$ are the same as those in Eq. (2.27), and $f^{eq}(t)$ and $f^{fr}(t)$ are the terms related to the evolution of the equilibrium state $g(\mathbf{x}, t, \mathbf{u})$ and initial distribution function $f_0(\mathbf{x}, \mathbf{u})$ with free transport, respectively. The above time-dependent solution f is similar to Eq. (2.26) in GKS, except for the initial distribution f_0. Once all terms in Eq. (3.11) are determined, the fluxes for the gas distribution function and conservative flow variables at the cell interface can be evaluated as

$$\mathcal{F}_f = \int_0^{\Delta t} \mathbf{u} \cdot \mathbf{n} f(\mathbf{x_0}, t, \mathbf{u}) ds dt$$

$$= \mathbf{u} \cdot \mathbf{n} S(q_1 \tilde{g}_0 + q_2 \tilde{g}_\mathbf{x} \cdot \mathbf{u} + q_3 \tilde{g}_t) + \mathbf{u} \cdot \mathbf{n} S \left(q_4 \tilde{f}_0 + q_5 \tilde{f}_{0\mathbf{x}} \cdot \mathbf{u} \right) \qquad (3.12)$$

$$= \mathcal{F}^{eq} + \mathcal{F}^{fr};$$

and

$$\mathbf{F_W} = \int_0^{\Delta t} \int \psi \mathbf{u} \cdot \mathbf{n} f(\mathbf{x_0}, t, \mathbf{u}) du ds dt$$

$$= \int \mathbf{u} \cdot \mathbf{n} S(q_1 \tilde{g}_0 + q_2 \tilde{g}_\mathbf{x} \cdot \mathbf{u} + q_3 \tilde{g}_t) \psi du + \int \mathbf{u} \cdot \mathbf{n} S \left(q_4 \tilde{f}_0 + q_5 \tilde{f}_{0\mathbf{x}} \cdot \mathbf{u} \right) \psi du$$

$$= \mathbf{F_W}^{eq} + \mathbf{F_W}^{fr},$$

$$(3.13)$$

where S is the area of the cell interface and $\mathbf{n}$ is its outward normal direction. The moments related to the Maxwellian function can be calculated analytically, and the moments of the initial distribution function are evaluated based on the numerical quadrature. The coefficients $q_1, q_2, \ldots, q_5$ are the same as those in Eq. (2.30) in the GKS fluxes. Here the equilibrium flux $\mathcal{F}^{eq}$ and the free transport flux $\mathcal{F}^{fr}$ at each specific particle velocity $\mathbf{u}_k$ come from the solution $f(\mathbf{x_0}, t, \mathbf{u})$ in Eq. (3.11), respectively.

The UGKS updates the conservative flow variables in Eq. (3.5) and the gas distribution function in Eq. (3.4) in a control volume in the following way,

$$\mathbf{W}^{n+1} = \mathbf{W}^n - \frac{1}{\Omega_\mathbf{x}} \sum_s \mathbf{F_W},$$

$$f^{n+1} = \left(1 + \frac{\Delta t}{2\tau^{n+1}} \right)^{-1} \left[f^n - \frac{1}{\Omega_\mathbf{x}} \sum \mathcal{F}_f + \frac{\Delta t}{2} \left(\frac{f^{+,n} - f^n}{\tau^n} + \frac{f^{+(n+1)}}{\tau^{n+1}} \right) \right],$$

$$(3.14)$$

where the interface fluxes for both macroscopic flow variables and microscopic distribution function are provided in Eq. (3.12)–(3.13) and the trapezoidal rule is used for the discretization of inner cell particle collision. After updating the macroscopic flow variables, the equilibrium state f^+ can be obtained. Then, the update of the velocity distribution function becomes straightforward. The update of the macroscopic flow variables in UGKS for a multi-scale solution is similar to that in GKS for the NS solution, except with a different initial gas distribution function f_0. In the GKS, f_0 is reconstructed using a Chapman-Enskog expansion from macroscopic flow variables. However, it is directly updated in UGKS at each particle velocity point in Eq. (3.14). In the rarefied flow regime, the direct update of f is necessary so as to capture the peculiar non-equilibrium state. In the continuum flow regime, especially when $\Delta t \gg \tau$, the update of macroscopic flow variables plays a dominant role for the flow evolution, because in the interface distribution function of Eq. (3.6), the contribution from f_0 decays exponentially as $\exp(-\Delta t/\tau)$, and the initial distribution function f_0 will not contribute much to the final solution. This should not be surprising, because in the continuum flow regime, the GKS is based on the macroscopic flow variables only. So, for continuum flow computation, the use of macroscopic flow variables is adequate for the design of the NS solver. The UGKS gets back to the GKS for NS solution in the continuum regime. It will be difficult to recover the NS solution with these kinetic solvers if only the gas distribution function f is updated. Even with the update of macroscopic flow variables in the implicit-explicit (IMEX) framework (Pieraccini & Puppo 2007; Dimarco & Pareschi 2013; Filbet & Jin 2010), if the interface flux is evaluated from the free transport of f_0 alone through the collisionless approach without including the contribution of the equilibrium state g, the scheme will not be accurate for obtaining the NS solution in the continuum flow regime once the cell size is considerably larger than the particle mean free path. As noticed in the integral solution (Eq. (3.6)), the determination of f will come from the equilibrium state mainly in the continuum flow regime, which is fully determined by macroscopic flow variables.

3.4 The Multi-scale Nature of UGKS

Many kinetic solvers have been developed in the past decades. A particular challenge for the kinetic method is to capture the hydrodynamic solution in the continuum limit without enforcing kinetic scale resolution, such as the automatic recovery of the NS solver. At the current stage, asymptotic preserving (AP) kinetic methods, which keep the same algorithm in different flow regimes, have been proposed, starting with the numerical method for radiative transfer (Larsen et al. 1987). However, the quality of the kinetic scheme under the AP analysis is indistinguishable, because it never presents the order for recovering the hydrodynamic equations. On the other hand, for a kinetic method, the accuracy in the transition regime depends closely on its limiting solution in the continuum limit under hydrodynamic scale resolution. In recent years, various kinetic schemes with AP properties have been developed for both the Euler and Navier-Stokes limits. Although all of the AP schemes can reach the Euler solutions in the limit of $Kn \rightarrow 0$, their asymptotic behaviours at the Navier-Stokes level are not clearly identified, even for those targeting the Navier-Stokes equations. In reality, AP schemes can have quite different performances in terms of viscous flow computations, due to their uncontrollable numerical dissipation from the free transport mechanism in the flux evaluation.

The UGKS provides a framework to study the flow of physics in a discretized space and time. The capturing of the multiple scale flow dynamics is based on the modelling of coupled particle transport and collision within a time step. The flow physics in UGKS is determined by the cell Knudsen number $Kn_C = \tau/\Delta t \sim \ell/\Delta x$. The integral solution (Eq. (3.6)) covers the flow regime from free molecular transport $Kn_C \geq 1$ to the Navier-Stokes wave propagation $Kn_C \ll 1$. The specific dynamics used in the local gas evolution depends on Kn_C. This direct modelling principle distinguishes the UGKS from many other kinetic solvers. Without using the integral solution or using the equivalent mechanism of direct coupling of kinetic solver and macroscopic governing equations (Yang et al. 2018; Su et al. 2020), any attempt to get an NS solution without imposing kinetic scale cell resolution will not be successful.

In order to make a scheme be able to recover the NS solutions in the continuum flow regime, it must be able to recover the NS Chapman-Enskog distribution function. More specifically, the dissipative term $-\tau(g_t + \mathbf{u} \cdot \nabla g)$ has to be recovered and used in the flux evaluation. However, many AP schemes use $f_0(\mathbf{x} - \mathbf{u}t)$ for the flux transport, and they can recover the Euler limit only in the continuum regime. In practice, it is meaningless to recover the Euler limit in the continuum regime for a multi-scale method because any conservative scheme can have such a property. Also, for a valid multi-scale method, the NS distribution function must be obtained under the condition of $\Delta t \gg \tau$. In order to clearly understand the property of the UGKS in the capturing of the NS solution, let's analyse the time-averaged UGKS flux of Eq. (3.11) or Eq. (2.29) in the x-direction:

$$\mathcal{F}_{UGKS} = \frac{1}{\Delta t} \int_0^{\Delta t} uf(\mathbf{x}_0, t, \mathbf{u})dt.$$

The main difference between the IMEX AP and the UGKS is the construction of the numerical flux.

The UGKS couples the collision and convection through a local evolution solution. When the cell Knudsen number Kn_C approaches zero, namely $\Delta t/\tau \gg 1$, the numerical flux in Eq. (3.11) reads:

$$\mathcal{F}_{UGKS} = u[1 - \tau(\bar{a}u + \bar{b}v + \bar{c}w + \bar{A}) + \frac{1}{2}\Delta t\bar{A}]g_0$$
$$+ \frac{\tau}{\Delta t}u[\mathrm{H}(u)f_0^l + (1 - \mathrm{H}(u))f_0^r - g_0] + O(\tau^2).$$
$$\text{(3.15)}$$

The first part is identical to Eq. (2.35) of the GKS NS flux in the smooth flow region. However, the flux used in most DVM methods, such as IMEX, will be the particle-free transport only:

$$\mathcal{F}_{DVM} = H(u)f_0^l + (1 - H(u))f_0^r.$$

As analysed in Eq. (2.37), the free transport term generates a numerical dissipation proportional to $\Delta t/2$, even though $f_0^{l,r}$ has the exact NS Chapman-Enskog distribution. Fortunately, in the UGKS, this free transport term is suppressed by g_0 first; then it is further reduced by a factor of $(\tau/\Delta t)$ in the continuum flow regime. Therefore, the numerical dissipation from the free transport in the UGKS diminishes quickly when $\Delta t \gg \tau$. The leading term in the UGKS flux is consistent with the NS solver, which is solely contributed from the integral part of the equilibrium state in Eq. (3.6), the same as that in the GKS. Therefore, the UGKS can preserve NS AP property with the hydrodynamic cell resolution. In fact, in the continuum limit, the collision term is considerably stiff. No matter what kind of initial distribution function f_0 is, the final distribution function will approach to a local equilibrium rapidly in the gas evolution process, and the equilibrium state will be fully determined by the macroscopic flow variables. This property is key for the UGKS to get back to the NS solver automatically in the continuum flow regime.

Although all AP schemes are able to obtain the Euler limit, only a few studies target the AP property of NS solutions. A valid multi-scale method should be able to go from kinetic to the Navier–Stokes solver with the decreasing of a cell's Knudsen number. Here, we would like to clarify some misunderstanding in the AP community. One of the claims is that one cannot expect to take $\Delta t \gg \tau, \Delta x \gg \ell$ and still capture the solution of the diffusion limit of the compressible Navier-stokes solution. It seems that the artificial dissipation, which is proportional to Δx or Δt due to the free transport in the kinetic solver, cannot be eliminated. The above analysis of the UGKS clearly indicates that the numerical dissipation on the order Δt in the continuum flow regime can indeed be eliminated by including particle collision in the transport process. In other words, in the continuum flow regime, the evolution of the equilibrium state plays a dominant role instead of the free transport of the initial non-equilibrium distribution. At the same time, there are different kinds of NS solutions in the continuum regime, and not all dissipative physical solutions have the scale of Knudsen number Kn or particle mean free path ℓ. For example, the Blasius boundary layer thickness of the Navier-Stokes equations is proportional to $\sqrt{\nu x/U} \sim \sqrt{Kn}$, and the thickness of the velocity profile can be hundreds of particle mean free paths ($\sim Kn$) at high Reynolds number cases. In the direct

NS solvers, it is a common practice to resolve the boundary layer solution with a few grid points, and each cell size can be comprised of dozens or hundreds of particle mean free paths. The shear layer thickness is similar in nature. Therefore, if the kinetic solver is claimed to go to the NS limit in the continuum flow regime, there is no reason why the standard NS solver cannot be fully recovered from the kinetic scheme. Similar to the GKS, the UGKS can capture the laminar boundary layer solution in the continuum flow regime with a few grid points. For particle-free transport or upwind-based kinetic solvers, it is hard to reach the boundary layer solution accurately, even though the cell size Δx is fine enough to resolve the boundary layer thickness $\sqrt{\ell}$, because its numerical dissipation ($\nu_n \sim \Delta t \sim \Delta x$) can be much larger than the physical one ($\nu_p \sim \ell$). This mechanism is similar to the flux vector splitting or kinetic flux vector splitting schemes for the Euler equations, in that the numerical dissipation will become dominant when $\Delta x \gg \ell$, due to the direct splitting of flux $F = F^+ + F^-$. These schemes require $\Delta x \leq \ell$ in order to properly control the numerical dissipation. On the other hand, the shock wave has a physical thickness of $O(\text{Kn})$. Similar to other shock-capturing schemes for the Euler and NS solutions, the UGKS cannot resolve the physical shock structure in the continuum flow regime and the numerical shock thickness is on the order of the cell size. For the direct NS solver, artificial numerical dissipation is added in the shock region explicitly or implicitly.

In order to evaluate the kinetic scheme at a small Knudsen number limit, recently a new concept of unified preserving (UP) has been proposed, which is used to assess the order of asymptotic limiting equations of the scheme (Guo et al. 2020; Liu & Xu 2020a). On the first-order approximation, this concept of UP is consistent with AP. But, on the second order, the UP analysis can distinguish the Navier-Stokes limit from the Euler one at a small Knudsen number, such as the recovery of the Chapman-Enskog expansion of the NS solution. In the end, it should be emphasized again that the UGKS method cannot capture a physical solution that cannot be resolved by the mesh size. In the unresolved case, the underlying governing equations will be changed with the variation of mesh size (Liu et al. 2019b). Instead, the UGKS method can only capture the NS solution once the thickness of the physical structure can be resolved by the mesh size, and the scheme does not require $\Delta x < \ell$ for the NS solutions, such as the computation of the laminar boundary layer and shear layer with thickness $\sim \sqrt{\ell}$. Once a kinetic scheme has a UP property and recovers the Chapman-Enskog NS distribution function under the condition $\Delta x \gg \ell$, it has a better chance for providing a physically reliable solution in the transition regime, especially in the near continuum flow simulation.

3.5 UGKS Validation

In the following subsection, a few examples will be presented to show the numerical performance of the UGKS.

(a) Lid-Driven Cavity Flow

In order to validate the multi-scale property of the UGKS, the cavity flow is studied in different flow regimes from transition to continuum. In the following calculations, the medium consists of argon gas modelled by a variable hard sphere. The wall temperature is kept the same as the reference temperature, $T_w = T_0 = 273K$, and the up wall velocity is kept fixed at $U_w = 50m/s$. Maxwell's diffusive boundary condition with full accommodation is used at all solid boundaries.

Based on the incoming particle distribution function f^{in} towards the boundary, which is obtained from the reconstruction of the initial distribution function next to the boundary, the outgoing particles from the boundary are assumed to have a Maxwellian distribution function g_w, where g_w has a wall temperature and velocity but the density is fully determined by the no particle penetrating wall condition $\int_{u>0} u g_w d\mathbf{u} + \int_{u<0} u f^{in} d\mathbf{u} = 0$.

The first case is for the rarefied flow at Knudsen number $Kn = 1$ and the UGKS solution is compared with the DSMC result (John et al. 2011; Liu 2016). The computational domain is composed of 50×50 non-uniform mesh in the physical space and 72×72 points in the velocity space. Figure 4 shows the results from the UGKS and the DSMC. The next test is to validate the UGKS solution in the continuum flow regime at Knudsen number $Kn = 1.42 \times 10^{-4}$ or $Re = 1000$. The computational domain is composed of 61×61 non-uniform mesh points in physical space and 32×32 points in the velocity space. Figure 5 shows the UGKS results and reference Navier-Stokes solutions (Ghia et al. 1982). This clearly demonstrates that the UGKS obtains the Navier-Stokes solutions accurately in the hydrodynamic limit, even with the cell size being much larger than the particle mean free path.

The UGKS can give accurate solutions in the rarefied and continuum flow regimes automatically. In the near continuum regime, it will be interesting to compare the results of the UGKS and the GKS. The GKS provides the NS solutions. In the following, the cavity case at $Re = 5$ or the Knudsen number $Kn = 2.85 \times 10^{-2}$ is tested. The results are shown in Figure 6. At this Reynolds number, the velocity profiles of the UGKS and the GKS (NS) are basically the same. However, even with a similar temperature distribution, the heat fluxes from the UGKS and the GKS are very different. The heat flux can be transported from a cold to a hot region in the UGKS due to the non-equilibrium dynamics beyond NS. The UGKS provides a more reliable physical solution than the Navier-Stokes equations. At the same time, at such a low Reynolds number, the time step used in the UGKS is solely determined by the CFL condition (Xu & Liu 2017), such as

$$\Delta t \sim \Delta x / \max(|\mathbf{U}| + \sqrt{RT}),$$

which is much larger than the traditional time step in the NS solver, such as the time step used in the GKS,

$$\Delta t \sim \min(\Delta x / \max(|\mathbf{U}| + \sqrt{RT}), (\Delta x)^2 / \nu),$$

where $(\Delta x)^2 / \nu$ is a constraint from the diffusion term in the NS formulation, which may indicate a weakness in the NS modelling for the low Reynolds number flow. In UGKS, there is no difference in the determination of the time step between high and low Reynolds number flows, because $(|\mathbf{U}| + a\sqrt{RT})$ is the largest physical speed for particle transport or wave propagation in the determination of the domain of dependence. The infinite propagating speed from the parabolic terms in the NS equations represents a weakness in the modelling, at least at the low Reynolds number flow limit.

(b) Flow Passing through a Circular Cylinder

It is challenging for any DVM kinetic solver to simulate hypersonic flow. Theoretically, the particle velocity space is infinite, and finite bounds must be chosen such that only a negligible fraction of molecules lies outside the computational domain. For hypersonic flow, due to the wide range of the velocity distribution, it is difficult to set a proper bound. A small fraction of molecules with very high velocity relative to the averaged bulk velocity have a significant effect on the macroscopic flow properties. As pointed out in (Bird 1994), if 100 grid points are employed in each direction in the velocity and physical space, the Boltzmann solver for a three-dimensional unsteady flow computation requires 10^{14} grid points. Therefore, the use of adaptive mesh in the velocity space becomes a preferred choice for the kinetic solver in the hypersonic flow simulation. If the grid in the particle velocity space can be automatically adapted to where it is needed, such as the particles in the DSMC method, it is possible that a kinetic solver can be used in a hypersonic rarefied flow study. In order to demonstrate the capacity of the UGKS with the adaptive velocity method in the hypersonic flow computation (Chen et al. 2012; Yu

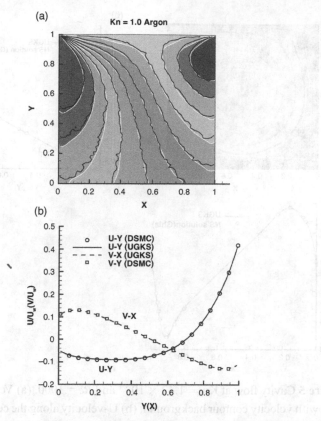

Figure 4 Cavity flow at Kn = 1. (a) Temperature contours, black lines: DSMC, white lines and background: UGKS; (b) *U*-velocity along the central vertical line and *V*-velocity along the central horizontal line, circles: DSMC, lines: UGKS. Courtesy of C. Liu (Liu 2016).

2013), the hypersonic rarefied flow of argon gas around a circular cylinder is tested. The UGKS results will be compared with the DSMC solutions.

The inflow has a Mach number $Ma = 20$ and Knudsen number $Kn = 1.0$. The Knudsen number is defined as the ratio of the particle mean free path over the radius of the cylinder. The hard-sphere model is used for defining the particle mean free path ℓ,

$$\ell = \frac{1}{\sqrt{2}\pi d^2 n},$$

where d is the molecular diameter and n is the number density. The density ρ_∞ is evaluated by $\rho_\infty = mn$, in which m is the molecular mass. The Mach number is defined as the ratio of the inflow velocity over the sound speed, i.e.,

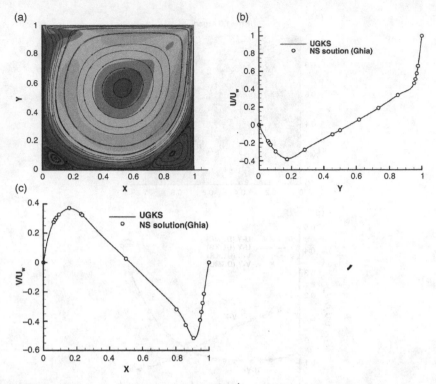

Figure 5 Cavity flow at Kn $= 1.42 \times 10^{-4}$ and $Re = 1000$. (a) Velocity stream lines with velocity contour background; (b) U-velocity along the central vertical line; (c) V-velocity along the central horizontal line. Circles: NS solution, lines: UGKS. Courtesy of C. Liu (Liu 2016).

$Ma = U_\infty/C_\infty$, where $C_\infty = \sqrt{\gamma R T_\infty}$ is the sound speed, γ is the ratio of specific heats, $R = k/m_c$ is the gas constant, and k is the Boltzmann constant. For an argon molecule, d is equal to $4.17 \times 10^{-10} m$, the molecular mass m_c is equal to $6.63 \times 10^{-26} kg$, and γ is equal to 5/3. The reference temperature T_∞ and the temperature at the cylinder surface T_w are set as constants, i.e., $T_\infty = T_w = 273K$. The viscosity is calculated using the variable hard sphere (VHS) model in DSMC

$$\mu = \mu_{ref} \left(\frac{T}{T_{ref}} \right)^\omega,$$

$$\mu_{ref} = \frac{15\sqrt{\pi \, m_c k T_\infty}}{2\pi d^2 (5 - 2\omega)(7 - 2\omega)},$$

where μ is the dynamic viscosity coefficient, T is the temperature, ω is the VHS temperature exponent, and $\omega = 0.81$. For the UGKS method, the reference

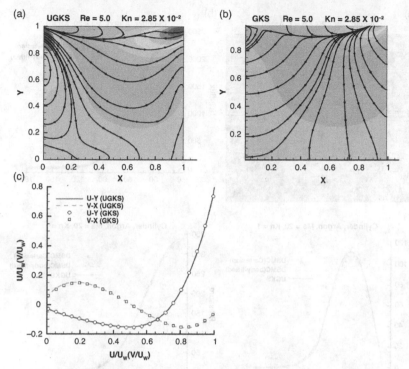

Figure 6 Cavity simulation using UGKS and GKS at $Kn = 2.85 \times 10^{-2}$ and $Re = 5$. (a) Temperature contour and heat flux: UGKS; (b) Temperature contour and heat flux of NS solutions: GKS; (c) U-velocity along the central vertical line and V-velocity along the central horizontal line, circles: GKS, lines: UGKS. Courtesy of C. Liu (Liu 2016).

density and temperature are set as ρ_∞ and T_∞, respectively. The initial condition is set with the inflow free stream parameters, while the initial velocity distribution is given by a Maxwellian function. The radius of the cylinder is set as $R = 0.01m$, and the computational domain is covered by both body-fitted and Cartesian meshes for DSMC and quadrilateral mesh for the UGKS. The mesh size for DSMC method is less than the particle mean free path, while there is no such limitation on the mesh size of the UGKS. The temperature distributions are shown in Figure 7a, where due to the use of the Shakhov model, the temperature rises a little bit earlier in the UGKS compared to that in the DSMC. This problem can be fixed using the unified gas-kinetic wave-particle methods provided in the next section. The surface properties around the cylinder, such as the pressure, shear stress, and heat flux, are presented in Figure 7. The UGKS and DSMC solutions match very well.

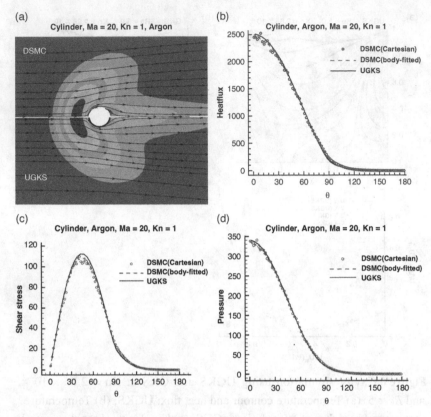

Figure 7 Flow passing through a cylinder at Ma = 20 and Kn = 1.0. (a) Temperature and stream line distributions (b) Heat flux; (c) Shear stress; (d) Pressure distributions along the cylinder surface. Courtesy of P.B. Yu (Yu 2013).

3.6 Summary

In this section, we present the framework for the construction of a unified algorithm for rarefied and continuum flow simulation. The UGKS is based on direct modelling in the discretized space. The mesh size and time step are used as the modelling scales for the flow transport. The dynamics of the flux evaluation at a cell interface depend on the cell's Knudsen number explicitly. Both the gas distribution function and the macroscopic flow variables are updated in the UGKS. The time step used in UGKS is not limited by the particle collision time. Once the time step Δt is a few times that of the particle collision time, the solution will not be sensitive to the individual binary particle collision and the flux in the UGKS will account for the accumulating effect of multiple particle collisions within a time step. Extensive tests from rarefied to continuum flow have been obtained. Even though the dynamics of the UGKS depend on the mesh size and time step, once the physical solution is well-resolved by the mesh size, further mesh refinement will no longer change the UGKS solution (Xu & Liu 2017; Liu et al. 2019b). For modest and low-speed rarefied flow, UGKS becomes much more efficient than DSMC, especially when the implicit and multigrid techniques are implemented (Zhu et al. 2017a; Zhu et al. 2017b; Zhu et al. 2019a). The UGKS method for diatomic gas with rotational and vibrational modes has been constructed as well (Yu 2013; Liu at al. 2014; Wang et al. 2017). The framework has also been extended to solve other multi-scale transport problems, such as radiative transfer, plasma, and gas-particle systems (Sun et al. 2015a, 2105b, 2017; Liu & Xu 2017; Liu et al. 2019a).

For high-speed and high-temperature rarefied flows, similar to other DVM examples, the UGKS method needs to use a large number of grid points in the particle velocity space, and then the scheme becomes expensive in terms of computational costs and memory requirements compared with the DSMC method. Even for the continuum flow simulation, the UGKS still uses the discrete particle velocity point in its DVM formulation when it gets back to the GKS. In order to reduce the computational cost for hypersonic flow simulation and remove the particle velocity space in the continuum flow regime, in the next section, we will introduce the particle and wave-particle version of the UGKS. At the same time, due to the use of particles, the original single relaxation kinetic model in the construction of the UGKS can be further improved by modifying the particle collision time according to its velocity. In other words, the UGKS can be constructed beyond the use of a single relaxation time kinetic model. The early temperature rise depicted in Figure 7a can be fully removed (Xu et al. 2021).

4 Unified Gas-Kinetic Particle and Wave-Particle Methods

4.1 Preface

In the previous sections, we presented the GKS method for the NS solution and the UGKS method for multi-scale flow transport. The GKS method has similar efficiency as a standard NS solver with the use of a continuous particle velocity space and analytical formulation in the flux evaluation in the update of macroscopic flow variables. In the GKS method, the gas distribution function at the beginning of each time step is reconstructed from macroscopic flow variables through the Chapman-Enskog expansion. The UGKS method is based on direct modelling in a discretized space for the construction of a unified algorithm. Like DVM, the UGKS method uses discrete particle velocity space. The conservative flow variables are updated, along with the gas distribution function.

The UGKS is an efficient method for low- and modest-speed non-equilibrium flow simulations due to an absence of noise and affordable number of grid points in the particle velocity space. The implicit and multigrid techniques can be naturally implemented in the UGKS to further improve the efficiency. In comparison with other DVM methods for the rarefied flow, the UGKS has distinguishable features. The time step used in the UGKS is not limited by the particle collision time, and the NS solution in the continuum flow regime can be automatically recovered. However, for hypersonic flow simulation, in order to cover a wide range of particle velocity space with discretized velocity points, the UGKS encounters great difficulties in terms of memory requirements and computational costs for updating the gas distribution function. In order to reduce the computational demand in the hypersonic flow computation and to get back to the exact GKS in the continuum flow simulation without using discrete particle velocity points, the unified gas-kinetic particle (UGKP) method and unified gas-kinetic wave-particle (UGKWP) method have been constructed for both monatomic and diatomic gas (Liu et al. 2020; Zhu et al. 2019b; Xu et al. 2020). In the UGKWP method, the update of the gas distribution function will be formulated using discrete particle and analytical wave. Instead of the finite volume formulation, the evolution of the combined particle and wave will be guided by the integral solution of the kinetic model equation directly. The distributions of the particles and waves depend on the cell's Knudsen number. In the continuum flow regime, the number of particles will diminish and the analytical NS distribution function is automatically obtained in the UGKWP, which recovers the GKS without using any particle. In the highly rarefied regime, the distribution function will be mainly represented by particles, and the efficiency of the particle method is retained in the UGKWP. This is a significant achievement in the development of an efficient multi-scale method for rarefied and continuum hypersonic flow simulations (Chen et al. 2020). Also, the UGKWP modelling is beyond continuum mechanics assumption, and it is an important step in the modelling and simulating of non-equilibrium physical process through the particle's non-local transport mechanism.

4.2 An Introduction to Particle Methods

The numerical methods for studying rarefied flow can be categorized into two groups: the stochastic particle method and the deterministic discretization method. For the stochastic method, the evolution of the velocity distribution function is represented by the motion of simulating particles. The direct simulation Monte Carlo (DSMC) method is the most popular and reliable method for the rarefied flow study (Bird 1994). Suffering from stochastic noise, the DSMC has low efficiency in low speed and small temperature variation flow simulation. At the same time, due to the separate physical processes of particle transport and collision, the numerical cell size and time step in the DSMC are constrained by the particle mean free path and particle collision time. Otherwise, as analysed in previous sections, the fact that the numerical dissipation is proportional to the time step cannot be ignored when the time step is larger than the particle collision time. Therefore, keeping the cell size smaller than the particle mean free path makes the DSMC highly expensive in the near continuum flow regime, such as the flow simulation for a vehicle at an altitude lower than 80 kilometres. In order to extend the DSMC to the continuum flow regime, the asymptotic preserving Monte Carlo methods (AP-DSMC; Pareschi & Russo 2000; Ren et al. 2014) have been constructed. However, once the decoupled particle transport and collision are adopted, this modified DSMC will not provide a reliable or accurate solution in the continuum regime. The single scale nature of the DSMC has not been changed much in the AP-DSMC methods. A laminar boundary layer calculation at a high Reynolds number becomes an outstanding test for validating these schemes. Among those stochastic kinetic methods, the particle Fokker-Planck method (Jenny et al. 2010) and the stochastic BGK method (Fei et al. 2020) target the multi-scale simulation over different flow regimes. The advantage of the particle method is its high computational efficiency in rarefied hypersonic flow simulation. The particle method is robust and not sensitive to mesh quality, but it suffers from statistical noise. In comparison with the particle method, the UGKS achieves high accuracy and efficiency in the near continuum flow regime. However, it suffers from the ray effect in highly rarefied flow computation (Zhu et al. 2020), and the computational cost is high for the hypersonic flow computation in any flow regime due to the discretized particle velocity space. In order to further improve the efficiency of the UGKS, the particle version will be presented here.

The UGKWP describes the gas distribution function and follows its evolution using both individual particle and analytic wave. During a time step, the trajectories of simulation particles are tracked until collision happens, and the post-collision particles are evolved collectively through the evolution of macroscopic flow variables which have a corresponding analytical formulation. The evolutions of simulation particles and analytical distribution functions are guided by the same time-dependent gas evolution model of the UGKS provided in Eq. (3.6). Similar to the UGKS, the UGKWP can provide accurate solutions in all regimes without requiring the cell size and time step to be less than the particle mean free path and collision time. The computational time and memory cost are on the same level as the particle methods in the highly rarefied regime and become comparable to the hydrodynamic flow solvers in the continuum regime. In fact, the UGKWP can get back to the GKS for the NS solutions without any particle in the continuum flow limit. The number of particles in the UGKWP is controlled dynamically by the cell's Knudsen number through the parameter $\exp(-\Delta t/\tau)$. Therefore, the UGKWP method will not have any stochastic noise in the continuum flow regime $\Delta t \gg \tau$.

The merit of UGKWP is the coupled evolution of waves and particles within a time step. In order to better explain the UGKWP, a particle version of the UGKS will be presented first, known as the UGKP method. Then, the UGKWP becomes a further optimization of the UGKP via the use of an analytic wave to represent collisional particles, and sampling collisionless particles only in the simulation. As a result, the UGKWP becomes an optimal method in a different regime, such as the particle one in the highly rarefied regime ($\Delta t \leq \tau$) and NS solver (GKS) in the continuum region ($\Delta t \gg \tau$). Between these two limits, a dynamic adaptation between waves and particles is obtained automatically according to the cell's Knudsen number. In other words, the local cell size and time step are the modelling scales for the development of a unified algorithm that controls the distribution and evolution of the local particles and waves.

4.3 The Unified Gas-Kinetic Particle Method

Particle tracking and interaction can be regarded as an optimal grid point adaption in the particle velocity space. The stochastic particle method obtains very high efficiency for simulations of high-speed rarefied flow in the three-dimensional case. In this section, the particle version of the UGKS will be presented in the construction of the UGKP method (Liu et al. 2020; Zhu et al. 2019b).

Within a time step Δt, one particle will keep free transport until it encounters another particle and collides. On average, a particle will take a free flight time τ before it collides with another particle. An individual particle will collide at time t_f, which can be much less than the numerical time step Δt. The free transport time t_f for each particle is determined statistically by the local particle collision time τ, which is a function of macroscopic flow variables. After each particle's first collision at t_f within a time step, the possible multiple or a massive number of collisions afterwards within the time interval ($t_f, \Delta t$) for these colliding particles will be tracked by the evolution of the equilibrium state through their contribution to the macroscopic flow variables. The relative value of local t_f and Δt identify the collisional or collisionless particle. The above description is a particle version for the evolution solution of Eq. (3.6).

Similar to the UGKS, the UGKP will update macroscopic flow variables

$$\mathbf{W}^{n+1} = \mathbf{W}^n - \frac{1}{\Omega_\mathbf{x}} \int_{t^n}^{t^{n+1}} \int_{\partial\Omega} \sum_k \psi \mathbf{u}_k \cdot \mathbf{n} f(\mathbf{x}, t, \mathbf{u}_k) ds dt, \tag{4.1}$$

where the integral solution (Eq. (3.6)) with a combination of analytical equilibrium state g and discrete particle for f_0 will be used in the determination of the above interface f and the flux calculation. Different from the DVM-type UGKS, the update of the gas distribution function will be based on the evolution of the particle in UGKP. The simulation particle $P_k(m_k, \mathbf{x}_k, \mathbf{u}_k, e_k)$ is represented by its weights of mass m_k, position $\mathbf{x}_k$, velocity $\mathbf{u}_k$, and internal energy e_k. In the UGKP, the same integral solution of the UGKS in Eq. (3.6) is used to capture the gas distribution function in $(\mathbf{x}, t)$ space with the initial particle distribution $f_0(\mathbf{x}, \mathbf{u})$,

$$f(\mathbf{x}, t, \mathbf{u}) = \frac{1}{\tau} \int_0^t e^{-(t-t')/\tau} g(\mathbf{x}', t', \mathbf{u}) dt' + e^{-t/\tau} f_0(\mathbf{x}_0, \mathbf{u}), \tag{4.2}$$

where the equilibrium distribution is integrated along the particle trajectory $\mathbf{x}' = \mathbf{x} + \mathbf{u}(t' - t)$, and $\mathbf{x}_0 = \mathbf{x} - \mathbf{u}t$ is the particle trajectory of the initial state. By substituting the expansion of equilibrium,

$$g(\mathbf{x}', t', \mathbf{u}) = g(\mathbf{x}, t, \mathbf{u}) + \nabla_\mathbf{x} g(\mathbf{x}, t, \mathbf{u}) \cdot (\mathbf{x}' - \mathbf{x}) + \partial_t g(\mathbf{x}, t, \mathbf{u})(t' - t)$$
$$+ O\left((\mathbf{x}' - \mathbf{x})^2, (t' - t)^2\right), \tag{4.3}$$

into the integral solution, the multi-scale evolution solution for the simulating particle is

$$f(\mathbf{x}, t, \mathbf{u}) = (1 - e^{-t/\tau}) g^+(\mathbf{x}, t, \mathbf{u}) + e^{-t/\tau} f_0(\mathbf{x}_0, \mathbf{u}), \tag{4.4}$$

where

$$g^+(\mathbf{x}, t, \mathbf{u}) = g(\mathbf{x}, t, \mathbf{u}) + \left(\frac{t e^{-t/\tau}}{1 - e^{-t/\tau}} - \tau\right)\left(\partial_t g(\mathbf{x}, t, \mathbf{u}) + \mathbf{u} \cdot \nabla_\mathbf{x} g(\mathbf{x}, t, \mathbf{u})\right). \tag{4.5}$$

The particle evolution solution (Eq. (4.4)) states that the probability of one particle experiencing no collision in a time interval $[0, t]$ is $e^{-t/\tau}$, and the colliding particle follows a velocity distribution g^+. The cumulative distribution function for particle's free streaming time t_f is

$$F(t \leq t_f) = \exp(-t_f/\tau), \tag{4.6}$$

from which t_f can be sampled from $t_f = -\tau \ln(\eta)$, where η is a uniform distribution $\eta \sim U(0, 1)$. For a time step Δt, the particle with $t_f \geq \Delta t$ is known as a collisionless particle and is retained in a time step, and the particle with $t_f < \Delta t$ is known as a hydrodynamic collisional particle and is eliminated in a time step. The total conservative quantities of collisionless particles in cell Ω_i will be defined by $\mathbf{W}_i^p$ by collecting all remaining particles in that cell after a one-time step, and the total conservative quantities of collisional or eliminated hydrodynamic particles in cell Ω_i can be evaluated by $\mathbf{W}_i^h$, which is the difference of $\mathbf{W}_i^{n+1}$ in Eq. (4.1) and $\mathbf{W}_i^p$, such that $\mathbf{W}_i^h = \mathbf{W}_i^{n+1} - \mathbf{W}_i^p$. Based on the conservative flow variables $\mathbf{W}_i^h$ and the distribution function g^+ for collisional particles in Eq. (4.5), these eliminated particles in the previous time step due to collision can be re-sampled at the beginning of the next time step. The detailed numerical procedure is provided as follows.

In the UGKP, the evolution of macroscopic flow variables in Eq. (4.1) follows the same evolution equation of the UGKS in Eqs. (3.12)–(3.14). The same equilibrium flux $\mathbf{F}_\mathbf{W}^{eq}$ in Eq. (3.13) of the UGKS will be used in the UGKP. The net free streaming flux $\mathbf{F}_\mathbf{W}^{fr}$ in Eq. (3.13) across the cell interface of the control volume Ω_i is calculated by counting all particles passing through the interface during a time step Δt,

$$\mathbf{F}_\mathbf{W}^{fr} = -\sum_{k \in P_{\partial\Omega_i^{in}}} \mathbf{W}_{P_k} + \sum_{k \in P_{\partial\Omega_i^{out}}} \mathbf{W}_{P_k}, \tag{4.7}$$

where $\mathbf{W}_{P_k} = \left(m_k, m_k\mathbf{u}_k, \frac{1}{2}m_k\mathbf{u}_k^2\right)$, $P_{\partial\Omega_\mathbf{x}^{out}}$ is the index set of the particles streaming out of the volume $\Omega_\mathbf{x}$ in the outward normal direction during a time step, and $P_{\partial\Omega_i^{in}}$ is the index set of the particles streaming into the volume $\Omega_\mathbf{x}$. Corresponding to the same terms of the UGKS in Eq. (3,14), the finite volume scheme for the update of conservative flow variables of the UGKP is

$$\mathbf{W}^{n+1} = \mathbf{W}^n - \frac{1}{\Omega_\mathbf{x}}\sum \mathbf{F}_\mathbf{W}^{eq} - \frac{1}{\Omega_\mathbf{x}}\mathbf{F}_\mathbf{W}^{fr}. \tag{4.8}$$

Here, the free transport flux $\mathbf{F}_\mathbf{W}^{fr}$ in Eq. (3.13) of the UGKS is replaced by real particle movement across the cell interface in Eq. (4.7).

The UGKP updates both macroscopic flow variables and discrete particles under the UGKS framework. For a time step from t^n to t^{n+1}, the algorithm of the UGKP method is summarized as follows (Liu et al. 2020; Zhu et al. 2019b):

(i) Stream the particles:

Based on the initial particle distribution function f_0, sample free streaming time $t_{f,k}$ for each particle P_k, and stream particle P_k for a time period of $\min(\Delta t, t_{f,k})$ to a location according to its velocity. Then, at the new location for each particle, the collisionless particle with $t_f \geq \Delta t$ is retained and the collisional particle with $t_f < \Delta t$ is eliminated.

(ii) Calculate the flux:

Calculate the net free streaming flow $\mathbf{F}_W^{fr}$ by counting the particles passing through the cell interfaces within a time step, and calculate the equilibrium flux $\mathbf{F}_W^{eq}$ in Eq. (3.13) from reconstructed macroscopic flow fields, as that in the UGKS.

(iii) Update the conservative variables:

Update total conservative variables $\mathbf{W}^{n+1}$ by Eq. (4.8). Sum up the total mass, momentum, and energy of all remaining collisionless particles $\mathbf{W}^{p,n+1}$ in the control volume, and calculate the conservative quantities of collisional or eliminated particles $\mathbf{W}^{h,n+1} = \mathbf{W}^{n+1} - \mathbf{W}^{p,n+1}$.

(iv) Update the particles:

Besides the collisionless particles in the control volume, based on the distribution g^+ determined by the updated $\mathbf{W}^{n+1}$, re-sample collisional particles with g^+ up to the number of particles with the total amount of flow quantities $\mathbf{W}^{h,n+1}$.

(v) Return to step (i) for the next time step computation.

The UGKP is a conservative finite volume method, where the particles are employed to recover the evolution solution of the non-equilibrium distribution function f. The macroscopic conservative flow variables $\mathbf{W}^{n+1}$ are also updated through the fluxes at the interfaces. On the microscopic level, all particles are tracked in the free streaming process for a time period $\min(\Delta t, t_f)$ and the collisional particles within a time step Δt are removed first; then all collisional particles within the control volume are re-sampled at time t^{n+1} from the equilibrium state g^+ with the total amount of macroscopic flow variables $\mathbf{W}^{h,n+1}$, where $\mathbf{W}^{h,n+1}$ is determined from the updated conservative flow variables $\mathbf{W}^{n+1}$ and those of the remaining particles $\mathbf{W}^{p,n+1}$, such as $\mathbf{W}^{h,n+1} = \mathbf{W}^{n+1} - \mathbf{W}^{p,n+1}$. Even when tracking and removing individual particles, the conservation law is precisely maintained in the above process, which is the key to the success of the current particle method. In addition, it should be noted that in the free transport process, each particle here moves over a free transport time t_f only, which is different from the DSMC particle for moving in a whole time step Δt. In other words, the DSMC particle always has $t_f = \Delta t$, which is different from the UGKP particle. With the consideration of the collisional process within a time step, the UGKP fully recovers the multi-scale nature of the UGKS.

4.4 The Unified Gas-Kinetic Wave-Particle Method

The UGKP uses particles to recover the evolution of the gas distribution function. According to Eq. (4.4), among the re-sampled particles from $\mathbf{W}^{h,n+1}$ at the beginning of the next time step, in the time step $\Delta t = t^{n+2} - t^{n+1}$, a proportion $(1 - \exp(-\Delta t/\tau))$ of the newly sampled particles will get collision. Therefore, not all re-sampled particles from $\mathbf{W}^{h,n+1}$ can survive in the next time step. Those collisional particles from $\mathbf{W}^{h,n+1}$ will again be deleted. As a result, it is not necessary to re-sample those collisional particles from $\mathbf{W}^h$, and their contributions can be tracked analytically. In the total $\mathbf{W}^h$, the number of collisional particles is calculated from

$$\mathbf{W}_c^h = (1 - e^{-\Delta t/\tau})\mathbf{W}^h,$$

and the number of collisionless particle is from

$$\mathbf{W}_p^h = e^{-\Delta t/\tau}\mathbf{W}^h.$$

Only the collisionless particles with the amount of $\mathbf{W}_p^h$ need to be sampled with free streaming time $t_f = \Delta t$. This optimizes UGKP and is called the UGKWP method. Here the gas distribution function is described by both the particle and wave. The wave part is referred to those collisional particles that can be tracked analytically.

In the UGKWP, the update of macroscopic flow variables is Eq. (4.8), and the particle transport part of $\mathbf{F}_\mathbf{W}^{fr}$ will be reformulated by taking into account the unsampled collisional particle from $\mathbf{W}_c^h$. The equilibrium flux $\mathbf{F}_\mathbf{W}^{eq}$ is still calculated from the macroscopic flow variables, which has the same formulation as the UGKP or UGKS. Again, the free streaming flux $\mathbf{F}_\mathbf{W}^{fr}$ in Eq. (4.8) represents the time integration of the particle free transport term $e^{-t/\tau}f_0(\mathbf{x}_0, \mathbf{u})$ in Eq. (4.4), which can be separated into two parts. The first part comprises the contribution from those particles retained from the previous time step and re-sampled collisionless particles from $\mathbf{W}_p^h$, which will follow the same particle transport of the UGKP by tracking their trajectories according to the time t_f. Their contribution to the flux $\mathbf{F}_\mathbf{W}^{fr}$ in Eq. (4.8) is denoted by $\mathbf{F}_{\mathbf{W}_p}^{fr}$. The second part comprises the contributions from the collisional particles from $\mathbf{W}_c^h$. These un-sampled collisional particles contribute to the $\mathbf{F}_\mathbf{W}^{fr}$ in Eq. (4.8) by the amount of $\mathbf{F}_{\mathbf{W}_c}^{fr}$, which can be obtained analytically.

The free streaming flux at the cell interface from the particles of $\mathbf{W}_c^h$, which is $\mathbf{W}_c^h = \mathbf{W}^h - \mathbf{W}_p^h$, is

$$
\begin{aligned}
\mathbf{F}_{\mathbf{W}_c^h}^{fr} &= \mathbf{F}_{UGKS}^{fr}(\mathbf{W}^h) - \mathbf{F}_{DVM}^{fr}(\mathbf{W}_p^h) \\
&= \int \mathbf{u} \cdot \mathbf{n}S\left(q_4 g_0^h + q_5 \mathbf{u} \cdot \nabla_x g_0^h\right)\psi d\mathbf{u} \\
&\quad - e^{-\Delta t/\tau} \int \mathbf{u} \cdot \mathbf{n}S\left(\Delta t g_0^h - \frac{1}{2}\Delta t^2 \mathbf{u} \cdot \nabla_x g_0^h\right)\psi d\mathbf{u},
\end{aligned}
\tag{4.9}
$$

where g_0^h is the Maxwellian determined by the total $\mathbf{W}$ for its temperature $\lambda = m/2kT$ and average velocity $\mathbf{U}$, but its density ρ is determined from $\mathbf{W}^h$. q_4, q_5 are given by Eq. (2.30). The flux $\mathbf{F}_{\mathbf{W}_p}^{fr}$ in $\mathbf{F}_{\mathbf{W}}^{fr}$ from particles passing through the cell interface is given by

$$
\mathbf{F}_{\mathbf{W}_p}^{fr} = -\sum_{k \in P_{\partial \Omega_i^+}} \mathbf{W}_{P_k} + \sum_{k \in P_{\partial \Omega_i^-}} \mathbf{W}_{P_k},
\tag{4.10}
$$

for cell Ω_i, where $\mathbf{W}_{P_k} = \left(m_k, m_k \mathbf{u}_k, \frac{1}{2}m_k \mathbf{u}_k^2\right)$, $P_{\partial \Omega_i^-}$ is the index set of the particles streaming out of the cell Ω_i during a time step, and $P_{\partial \Omega_i^+}$ is the index set of the particles streaming into the cell Ω_i. In comparison with Eq. (4.8) of the UGKP, the update of conservative flow variables of the UGKWP is

$$
\mathbf{W}_i^{n+1} = \mathbf{W}_i^n - \frac{1}{|\Omega_x|}\sum \mathbf{F}_{\mathbf{W}}^{eq} - \frac{1}{|\Omega_x|}\sum \mathbf{F}_{\mathbf{W}_c^h}^{fr} - \frac{1}{|\Omega_x|}\mathbf{F}_{\mathbf{W}_p}^{fr}.
\tag{4.11}
$$

The algorithm for the UGKWP is as follows. Similar to the UGKP, the main difference in the UGKWP is that only the collisionless particles from $\mathbf{W}_p^h$ (instead of $\mathbf{W}^h$) are sampled in the simulation, and the free streaming flux $\mathbf{F}_{\mathbf{W}_c^h}^{fr}$ from the non-sampled particles of $\mathbf{W}_c^h$ is computed analytically. A net particle flow $\mathbf{F}_{\mathbf{W}_p}^{fr}$ is composed of the contributions from all particles, which include the remaining particles from the previous time step and the newly sampled collisionless particles from $\mathbf{W}_p^h$. The numerical procedures of the UGKWP include:

(i) Stream the particles:

Assign $t_{f,k} = \Delta t$ for the re-sampled collisionless particles from $\mathbf{W}_p^h$. Sample free streaming time $t_{f,k}$ for each remaining particle from the previous time step. Stream each particle P_k for a time $\min(\Delta t, t_{f,k})$. Then, at the new location for each particle, the collisionless particles with $t_f \geq \Delta t$ are retained, and the collisional particles with $t_f < \Delta t$ are eliminated.

(ii) Calculate the flux:

Calculate the net free streaming flux $\mathbf{F}^{fr}_{\mathbf{W}_p}$ in Eq. (4.10) by counting the particles passing through the cell interfaces. Calculate the analytic free streaming flux $\mathbf{F}^{fr}_{\mathbf{W}^h_c}$ in Eq. (4.9) using the collisional particles of $\mathbf{W}^h_c$. Similar to the UGKP, based on the formulation in Eq. (3.13), calculate the equilibrium flux $\mathbf{F}^{eq}_{\mathbf{W}}$ in Eq. (4.11).

(iii) Update the conservative flow variables:

Update the total conservative flow variables $\mathbf{W}$ using Eq. (4.11). Calculate the flow variables $\mathbf{W}^p$ of collisionless particles by summing up the mass, momentum, and energy of all remaining particles in the cell. Evaluate the conservative variables of eliminated collisional particles,

$$\mathbf{W}^h = \mathbf{W} - \mathbf{W}^p.$$

(iv) Update the particles:

The particles now comprise the remaining particles from the previous time step and re-sampled collisionless particles from $\mathbf{W}^h$ with the free streaming time $t_f = \Delta t$ with a total amount of macroscopic flow variables $\mathbf{W}^h_p = e^{-\Delta t/\tau}\mathbf{W}^h$ with a corresponding equilibrium state g^+ determined by the total $\mathbf{W}$.

(v) Go to step (i) for next time step computation.

4.5 Analysis and Further Development of UGKWP

The gas distribution function in the UGKWP is represented by particles and waves. Their weights depend on the cell's Knudsen number. The UGKWP is a multi-scale method and is able to present an accurate solution for all flow regimes, which is especially suitable for the hypersonic flow computation. In the kinetic scale with $\Delta t \leq \tau$, the UGKWP becomes a particle method to track the evolution of the gas distribution function. In the continuum regime $\Delta t \gg \tau$, the UGKWP becomes a second-order GKS for the Navier-Stokes solutions and the flux is mainly from the evolution of the equilibrium state, which can be evaluated solely from the macroscopic flow variables. For the UGKS, even in the continuum flow regime, the discretized particle velocity space is still kept, which makes the scheme expensive. However, for the UGKWP, in the continuum flow regime, the particles will disappear and the full GKS with an analytic formulation is recovered. In the continuum regime as $\tau \to 0$, the time evolution of the gas distribution function in the UGKWP goes to

$$g^+(\mathbf{x}, t, \mathbf{u}) = g(\mathbf{x}, t, \mathbf{u}) + \left(\frac{te^{-t/\tau}}{1 - e^{-t/\tau}} - \tau\right)\left(\partial_t g(\mathbf{x}, t, \mathbf{u}) + \mathbf{u} \cdot \nabla_x g(\mathbf{x}, t, \mathbf{u})\right)$$

$$= g(\mathbf{x}, t, \mathbf{u}) - \tau\left(\partial_t g(\mathbf{x}, t, \mathbf{u}) + \mathbf{u} \cdot \nabla_x g(\mathbf{x}, t, \mathbf{u})\right) + O(e^{-\Delta t/\tau}),$$

$$(4.12)$$

which is exactly the same as the NS gas distribution function. The flux $\mathbf{F_W}$ for macroscopic flow variables of the UGKWP in Eq. (4.1) will become

$$\mathbf{F_W} = \Delta t \int \mathbf{u} \cdot \mathbf{n} S\left(g - \tau \mathbf{u} \cdot \nabla_x g + \left(\frac{1}{2}\Delta t - \tau\right)\partial_t g\right)\psi d\mathbf{u} + O(e^{-\Delta t/\tau}), \quad (4.13)$$

which recovers the GKS flux for the NS solution in Eq. (2.35) in the smooth flow region.

In the UGKWP, as $\tau \to 0$, the total mass of simulating particles will have $e^{-\Delta t/\tau}\rho^h\Omega_i \to 0$. Therefore, the statistical noise exponentially decays in the continuum regime due to the absence of particles. When a fixed mass m_p is used for simulating a particle, the number of total particles $N_P \to 0$ in the continuum flow regime. The UGKWP preserves the multi-scale property of the UGKS from the collisionless particle transport to the hydrodynamic Navier-Stokes solution, and the cell size and time step are not constrained by the particle mean free path and collision time. The computational efficiency of the UGKWP achieves its highest level in different regimes for hypersonic flow computation. The computational cost, including the computational time and memory requirement, is comparable to the NS solvers in the continuum limit. At the same time, for highly non-equilibrium hypersonic flow, the efficiency goes to the purely particle method. The UGKWP is a suitable method for high-speed flow simulation. For low-speed microflow, the UGKS still has an advantage due to the absence of noise, less grid points used in the velocity space, and the possible use of acceleration techniques.

In the highly rarefied flow regime, the UGKWP will use mainly the particles for the flow evolution. Similar to the UGKS, the particle evolution in the UGKWP is still controlled by the kinetic relaxation model, such as the Shakhov model. In other words, the collision time in the UGKWP for all particles has the same relaxation time regardless of how high or low the individual particle velocity is. This is definitely inconsistent with the physical reality, where the collision time τ or collision frequency should be related to the particle velocity $\tau = \ell/|\mathbf{u}|$. In order to incorporate the physical effect, a direct modification of the particle transport in the UGKWP can be performed for these relatively high-velocity particles (Xu et al. 2021). The newly modified τ^* has the form

$$\tau^* = \begin{cases} \tau & \text{if } |\mathbf{u}-\mathbf{U}| \leq 5\sigma \\ \dfrac{1}{1+0.1|\mathbf{u}-\mathbf{U}|/\sigma}\tau & \text{if } |\mathbf{u}-\mathbf{U}| > 5\sigma \end{cases} \tag{4.14}$$

where $\sigma = \sqrt{RT}$. Then, for the particle, the free streaming time is determined by $t_f = -\tau^*\ln(\eta)$.

For the high-speed particle, the relative collision time will be reduced according to the particle velocity. For other particles, the free streaming time remains the same, which has been taken into account properly by the kinetic relaxation model, such as that for recovering the viscosity and heat conduction coefficients. The above modification of particle collision time does not affect the conservation of the scheme. Even for the particles with modified τ, their mass, momentum, and energy will stay the same, and the only modification is the distance it travels. This modification of the particle free streaming time can be done directly in the UGKWP. For the DVM-based UGKS, it is very hard to modify the collision time at a particular particle velocity point and keep the overall conservation of the system. In other words, due to the adaptation of particle, more realistic kinetic modelling can be implemented directly into the UGKWP, which goes beyond the kinetic model equations with a single relaxation time.

4.6 Numerical Validation

In this section, the performance of the UGKWP will be presented using a few numerical examples.

(a) Shock Structure Calculation

One of the simplest and most fundamental non-equilibrium gas-dynamic phenomena that can be used for algorithm validation is the internal structure of a normal shock wave. There are two main reasons for this. First, the shock wave represents a flow condition that is far from thermodynamic equilibrium. Second, shock wave phenomena are unique in that they allow one to separate fluid dynamics from the boundary condition. The boundary condition for a shock wave is simply determined by the Rankine-Hugoniot relation. Thus, in the study of shock structure, one is able to identify the flow physics of a modelling scheme. Since 1950s, the computation of shock structure has played an important and critical role for validating kinetic theory and numerical schemes in non-equilibrium flow studies. In the following calculations, the viscous coefficient is given by

$$\mu = \mu_{ref}\left(\frac{T}{T_0}\right)^{\omega},$$

with the reference viscosity

$$\mu_{ref} = \frac{15\sqrt{\pi}}{2(5 - 2\omega)(7 - 2\omega)}\frac{\ell}{L},$$

where L is the characteristic length and ω is the temperature dependence index. The Mach 8 argon shock structure is calculated with $\mu \sim T^{0.68}$ and compared with the DSMC solution (Bird 1970). Figure 8 shows the distributions of density, temperature, heat flux, and stresses from the UGKWP calculation and the DSMC solutions. Here the particle collision time is defined by Eq. (4.14). In comparison with the Shakhov model–based UGKS result (Xu & Huang 2011), the solution from the current scheme is greatly improved.

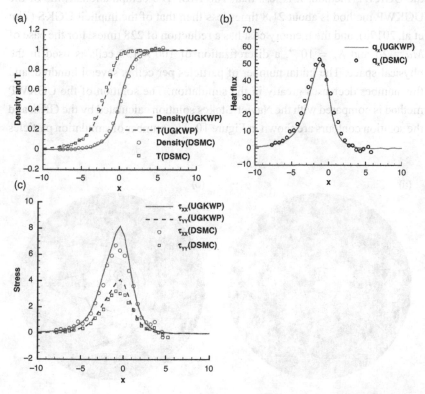

Figure 8 Mach 8 argon shock structure calculated by the UGKWP and the DSMC (Bird 1970). The x-coordinate is normalized by ℓ. (a) Density and temperature, (b) heat flux, (c) shear stress. Courtesy of X.C. Xu.

(b) Flow Passing through a Circular Cylinder

The hypersonic flow of argon gas passing over a circular cylinder at different Knudsen numbers is calculated by the UGKWP. For argon gas, the parameters are defined in the same way as the cylinder case in the UGKS calculation of the last section. This case has a Mach number $M_a = 20$ and a Knudsen number $K_n = 1.0$. The discretization of the physical domain is 64 cells along the azimuthal direction and 110 cells along the radial direction. In order to compare with the performance of the UGKS, the velocity space for the UGKS is covered on a range of $[-50,50] \times [-50,50]$, which is discretized by 200×200 velocity points. Figure 9 shows the contours of steady-state pressure and temperature distributions calculated by the UGKWP and the UGKS. The average number of simulation particles per cell in the UGKWP solution is shown in Figure 11a. To get a steady state solution, the computational time for implicit UGKS is about 13.1 hours, and the explicit UGKWP method takes 36.1 minutes (including the averaging procedure). The memory cost for the UGKS is 22.3 GB, and for the UGKWP method, it is less than 100 MB. The computational time of the UGKWP method is about 21.8 times less than that of the implicit UGKS (Zhu et al. 2017a), and the memory cost has a reduction of 228 times. For the case of $Ma = 20$ and $K_n = 10^{-4}$, a discretization of 100×150 cells is used in the physical space. The initial number of particles per cell is several hundred, and this number decreases greatly in the simulation. The solution of the UGKWP method is compared with the Navier-Stokes solution calculated by the GKS, and the solution contours are shown in Figure 10. The number of simulation particles

(a)   (b)

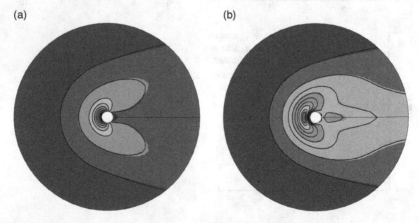

Figure 9 (a) Pressure and (b) temperature contour for $Ma = 20$ and $Kn = 1$. The UGKWP method solution is shown in flood, and the UGKS solution is shown in the contour line. Courtesy of C. Liu (Liu et al. 2020).

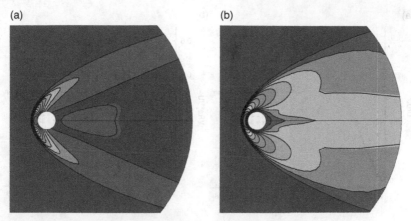

Figure 10 (a) Pressure and (b) temperature contour for $Ma = 20$ and $Kn = 10^{-4}$. The UGKWP method solution is shown in flood, and the GKS solution is shown in contour lines. Courtesy of C. Liu (Liu et al. 2020).

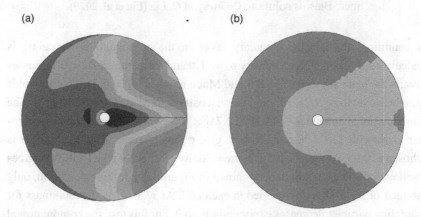

Figure 11 Number of simulation particles per cell for the cylinder flow with $Ma = 20$: (a) $Kn = 1.0$. Depending on region, there are 20 to 320 particles per cell; (b)$Kn = 1.0 \times 10^{-4}$. There are only 0 to 2 particles per cell. Courtesy of C. Liu (Liu et al. 2020).

per cell is shown in Figure 11b. The computation time for the UGKWP method is 17.2 minutes, which is comparable to the efficiency of a standard NS solver.

(c) Laminar Boundary Layer

It is challenging for a particle method to calculate the Navier-Stokes solution under a cell size that is much larger than the particle mean free path and a time step much larger than the particle collision time. To show the ability of the UGKWP method in capturing the Navier-Stokes solution under such

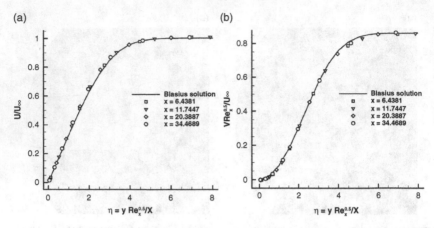

Figure 12 Velocity profile of a UGKWP method comparing to the Navier-Stokes Blasius solution. (a) U-velocity distribution at different locations; (b) V-velocity distribution at different locations. Symbols: solution of the UGKWP, lines: Blasius solution. Courtesy of C. Liu (Liu et al. 2020).

a condition, the laminar boundary layer in the continuum flow regime is calculated. A gas flow with density $\rho_0 = 1.0$ and temperature $T_0 = 0.11$ passes over a flat plate at speed $U_0 = 0.1$ and Mach number $M_a = 0.33$. The Reynolds number is set to be $R_e = 10^5$, and the viscosity is fixed at $\mu = 1.05 \times 10^{-4}$. The computational domain is $[-44.16, 112.75] \times [0, 29.8]$, and a stretched rectangular mesh with 120×30 non-uniform grid points is used. The CFL number is chosen as 0.95. At the steady state, the velocity profile from the UGKWP agrees well with the Blasius solution, as shown in Figure 12. For this calculation, only about 1 or 2 particles are retained in each cell. As $\tau/\Delta t \to 0$, the total mass for sampling particles decreases exponentially to 0. For this test, the computational time is less than 2 minutes, which is comparable to the efficiency of the standard Navier-Stokes solver.

(d) Flow Passing through a Space Vehicle

This is an example of Mach number 9 monatomic gas passing through a vehicle at Knudsen number $K_n = 10^{-3}$ with a characteristic length $0.28m$. The monatomic argon gas has molecular mass $m = 6.63 \times 10^{-26} kg$ and diameter $d = 4.17 \times 10^{-10} m$. The incoming gas with the viscosity coefficient index $\omega = 0.81$ has a temperature $56K$ and an angle of attack (AOA) 20° on the vehicle. There are 15,277 pyramids and 545,316 tetrahedrons around the vehicle, with a total of 560,593 cells. Due to the symmetry, only half of the vehicle is simulated. The boundary condition on the vehicle surface is

(a)

(b)

(c)

(d)

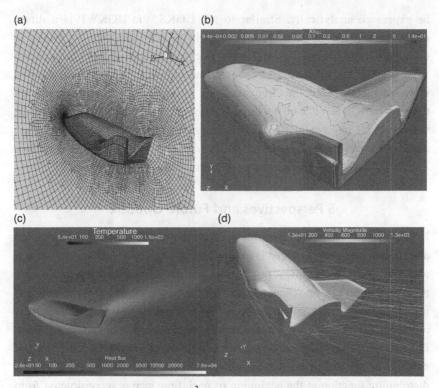

Figure 13 Mach = 9 and $Kn = 10^{-3}$ flow around a vehicle at AOA 20°. (a) Mesh around the vehicle; (b) local Knudsen number; (c) temperature distribution; (d) velocity and streaming lines. Courtesy of Y.P. Chen.

a diffusive one, on which the temperature maintains at $T_w = 300K$. The CFL number for the simulation is 0.95. The simulation runs 35.86 hours with 48 cores (each core has a 2.2 GHZ CPU) and consumes 58 GB of memory. The mesh, local Knudsen number, temperature, and velocity-streaming lines are shown in Figure 13.

4.7 Summary

In this section, the multi-scale unified gas-kinetic particle and wave-particle methods are presented. Based on the wave-particle decomposition for the gas distribution function, the UGKWP keeps the multi-scale property of the UGKS but abandons the discrete particle velocity space. As a result, for hypersonic flow computation, the UGKWP becomes a highly efficient method. The distinguishable feature of the UGKWP is that it will automatically get back to the GKS for the Navier-Stokes solutions in the continuum flow regime, where the flux transport from the gas distribution function can be expressed analytically. Similar to the UGKS, the UGKWP is a direct modelling scheme on the scales of the cell size and time step. According to the cell's Knudsen number, the scheme captures flow physics seamlessly from the Boltzmann particle transport and collision to the Navier-Stokes wave interaction. With the direct adoption of particles, the particle collision time in the UGKWP can be directly modelled according to the particle velocity. This implementation greatly improves the accuracy of the scheme, and the underlying kinetic model is beyond the single relaxation equation. The UGKWP has been extended to diatomic gas as well (Xu et al. 2020).

5 Perspectives and Future Outlook

In this Element, the unified computational fluid dynamics algorithm for rarefied and continuum flow simulation is introduced. In order to understand the central idea and grasp the historical progress in the development of the multi-scale method, the gas-kinetic scheme (GKS) for the Navier-Stokes (NS) solutions is introduced first, where only macroscopic flow variables are updated through the numerical fluxes, which are evaluated from the time-dependent gas distribution function at a cell interface. The evolution solution is based on the integral solution of the kinetic model equation, and the initial condition of the gas distribution function at the beginning of each time step is reconstructed from macroscopic flow variables through the Chapman-Enskog expansion. Even though the GKS is similar to the Godunov-type CFD methods targeting on the macroscopic Euler and NS equations, the gas evolution mechanism in the GKS and Riemann solver is distinguishable. The GKS uses a physical relaxation process from the initial non-equilibrium state to an equilibrium one in the determination of the flux function, and the flux covers a multi-scale process from the collisionless particle transport to the NS wave propagation. The GKS is basically a discretized physical conservation law, and the transport depends on the cell's Knudsen number. The NS solution is obtained when the flow structure is well resolved by the cell size which is much larger than the particle mean free path. For the unresolved shock discontinuity, the GKS introduces numerical dissipation by enlarging the cell's Knudsen number and incorporates the particle free transport in constructing a numerical shock structure with cell size thickness. For the Godunov method, the equilibrium state is always assumed everywhere, even inside the dissipative shock layer. The dissipation in the Godunov method is mainly provided in the preparation of the initial condition, and it depends sensitively on the relative orientation between the mesh and shock front. Due to the multidimensional property in the GKS flux function, the scheme becomes less sensitive to the mesh orientation and the evolving wave at a cell interface can propagate in any direction, starting from a multidimensional initial reconstruction with flow gradients in both the normal and tangential directions. Based on the time-accurate evolution model, the GKS becomes very efficient when implementing a two-stage fourth-order technique instead of the Runge-Kutta time-stepping method for the temporal evolution. Also, the time-dependent gas distribution function in the GKS can be used to update the cell interface values, which can be used to construct high-order compact GKS. The sixth- and eighth-order compact GKS provide excellent performance from the acoustic wave calculation to the high Mach number shock interaction (Zhao et al. 2019). The GKS becomes one of the most robust, accurate, and efficient CFD methods for the compressible flow simulation in the continuum flow regime (Zhao et al. 2020a, 2020b).

In order to extend GKS from a continuum flow solver to the rarefied flow regime, the evolution of the peculiar non-equilibrium gas distribution function has to be followed instead of reconstructed in the GKS. The unified gas-kinetic scheme (UGKS) was developed with the updates of both macroscopic flow variables and the gas distribution function. Similar to the DVM, the particle velocity space in the UGKS is discretized by grid points. Since the gas distribution function is updated, there is no need to use the Chapman-Enskog theory to reconstruct it at the beginning of each time step. In the UGKS, the evolutions of macroscopic flow variables and the gas distribution function are fully coupled. In terms of the flux evaluation, the UGKS and GKS use the same integral solution of the kinetic model equation, which covers the particle free transport and hydrodynamic wave propagation. The key ingredient in the UGKS is that the cell's Knudsen number determines the evolution solution used in the construction of the scheme. Therefore, the UGKS is also known as the direct modelling method. It uses the mesh size and time step as the modelling scales to construct governing equations directly in a discretized space. It is different from the traditional numerical partial differential equation (PDE) methodology, where the scheme is used to solve the pre-determined PDE. Depending on the cell's Knudsen number, the UGKS provides discretized governing equations and the flow physics in various regions can be different, such as the compressible NS solution at the leading edge and free molecular flow at the trailing edge in a high-speed flight simulation. The UGKS is not equivalent to a kinetic equation solver, such as the direct discretization of the Boltzmann equation. In the UGKS, the dynamic effect from multiple particle collisions within a time step is included in both the flux transport and inner cell collision. The coupled particle transport and collision within a time step is the key to preserve the multi-scale nature of the UGKS. Any scheme based on the decoupled particle transport and collision, such as the DSMC and direct Boltzmann solver, will encounter difficulties in the construction of a multi-scale method. For the time evolving upwind-based DVM, the scheme is only valid when the cell size is less than the particle mean free path, i.e., in the so-called kinetic scale resolution. Even though the GKS and the UGKS are based on the same evolution model in the flux evaluation, in the near continuum flow regime, the difference between updating the gas distribution function in the UGKS and reconstructing the gas distribution function in the GKS can cause distinguishable variance in the solution, such as the heat flux in the low Reynolds number flow simulation. At the same time, for low Reynolds number flow, the GKS has the same requirement in the determination of the time step as the standard NS solver, where besides satisfying the CFL condition $\Delta t \leq \Delta x / \max(|\mathbf{U}| + \sqrt{\gamma RT})$, the stability condition from the diffusion term imposes the requirement $\Delta t \leq (\Delta x)^2 / \nu$ as well. However, the time step in the UGKS is fully determined by the CFL condition. Physically, for a gas system, whether the Reynolds number is high or low, the maximum propagating speed from the particle movement or the collective wave propagation will be related to $|\mathbf{U}| + \sqrt{\gamma RT}$ only. Therefore, the constraint from $(\Delta x)^2 / \nu$ on the time step in the NS solver for low Reynolds number flow indicates the inappropriate physical modelling of the NS equations. For example, the Chapman-Enskog expansion is problematic at a low Reynolds number (Xu & Liu 2017). In the low Reynolds number limit, the particle penetration between fluid elements should be taken into account when the particle mean free path is not totally negligible in comparison with the size of the fluid element. The governing equations in the multi-scale UGKS are not based on the closed fluid element concept and have no continuum mechanics assumptions that are used in the reconstruction of hydrodynamic equations and all those moments equations. The UGKS presents an accurate solution for rarefied and continuum flow. For the hypersonic flow simulation, the large density variation can cause significant changes in the local Knudsen number, with a difference of several orders of magnitude around a vehicle in a near space flight, and the UGKS can capture such a multi-scale solution according to local cell's Knudsen number.

Like the discrete velocity method (DVM), the UGKS uses grid points in the particle velocity space to update the gas distribution function. For low and modest speed or near continuum flow regime, the velocity space can be well discretized by a reasonable number of grid points. Due to the absence of statistical noise and the implementation of implicit and multigrid techniques, the UGKS is a very efficient method for flow computation. For microflow at low speed, the UGKS is an outstanding method in terms of its accuracy, efficiency, and multi-scale nature. However, for high-speed flow, in order to cover the widespread particle velocity space and capture the large variation of the gas distribution function at extremely low and high temperatures, a large number of grid points have to be used to cover the whole particle velocity space. This increases the memory and computational cost significantly for hypersonic flow computation. Moreover, even in the continuum flow regime, the discrete particle velocity points are still kept in the update of the gas distribution function. The efficiency of the UGKS in the continuum flow regime can be hardly compared with the GKS and other standard NS solvers. At the same time, for the highly rarefied flow, as the same as other DVM methods, the ray effect may appear in the UGKS due to inadequate resolution in the particle velocity space. In order to improve the efficiency of the UGKS for high-speed flow, the scheme is further developed by incorporating particles and waves to represent the gas distribution function. Under the same UGKS framework, besides updating macroscopic flow variables inside each control volume, the gas distribution function can be tracked by particles in the unified gas-kinetic particle (UGKP) method.

As a multi-scale method, in UGKP the particle is categorized into collisionless and collisional particles within a time step according to the particle's free streaming time t_f. The trajectory of the collisionless particle within a time step is fully tracked, including its path across the cell interfaces. The collisional particle within a time step is tracked up to free streaming time t_f and removed. These collisional hydrodynamic particles get re-sampled from the equilibrium state at the beginning of the next time step from the updated macroscopic flow variables $\mathbf{W}^h$ of all eliminated collisional particles inside the control volume. With the updated total macroscopic flow variables $\mathbf{W}$ and those from all remaining particles $\mathbf{W}^p$, the macroscopic flow variables of the annihilated hydro-collisional particles are given by $\mathbf{W}^h = \mathbf{W} - \mathbf{W}^p$. Due to the use of particles, the statistical noise in the UGKP is inherent, even in the continuum flow regime.

In order to further improve the efficiency of the UGKP, the unified gas-kinetic wave-particle (UGKWP) method is developed. The main modification from the UGKP to the UGKWP is that not all particles from $\mathbf{W}^h$ need to be re-sampled, because a portion of these re-sampled particles would collide again in the next time step and be removed. Thus only collisionless particles need to be sampled, and the amount of macroscopic flow variables corresponding to collisionless particles for re-sampling equals $e^{-\Delta t/\tau}\mathbf{W}^h$. The transport of other collisional hydro-particles with the amount $(1 - e^{-\Delta t/\tau})\mathbf{W}^h$ can be evaluated analytically and treated as a wave. This is equivalent to using particles for the non-equilibrium transport only and using analytic waves for equilibrium transport. The dynamic coupling between the particle and wave depends on the cell's Knudsen number. In the highly rarefied flow regime $\Delta t \leq \tau$, similar to the UGKP, all particles will be retained and evolved. In the continuum flow regime $\Delta t \gg \tau$, there are almost no particles left from $e^{-\Delta t/\tau}\mathbf{W}^h$. In the transition regime, the portions of waves and particles depend on the local cell's Knudsen number $Kn_c = \tau/\Delta t$, and are proportional to $(1 - e^{-1/Kn_c})$ and e^{-1/Kn_c}. Therefore, in the continuum flow regime, almost no particles will appear and the UGKWP will automatically go back to the GKS for the NS solution. In terms of physical modelling, the UGKWP gives an optimal description of flow physics in different regimes. The particle is used to capture non-equilibrium transport. The flexible number of particles naturally provides the adaptive degrees of freedom needed for the non-equilibrium flow description. Furthermore, due to the use of particles, the collision time for individual particle can be modified according to particle velocity. As a result, even starting from the kinetic model equation, the UGKWP can include the flow physics, which is beyond the kinetic model with a single relaxation time. Also, due to the introduction of particle the ray effect in UGKS will not appear in UGKWP.

The success of the UGKS is due to the direct modelling methodology. The flow dynamics corresponding to the local cell's Knudsen number is used in the construction of the algorithm. The mesh size and time step play dynamic roles for the description of flow evolution. With the variation of mesh size and time step, the accumulating effect from particle transport and collision up to the time step scale is used for the flow evolution and flux evaluation. In other words, the UGKS uses the cell size and time step as the modelling scales to construct the governing equations. The flow physics in the UGKS covers the kinetic scale Boltzmann equation and the hydrodynamic scale NS equations, presenting a unified framework to connect them seamlessly. The particle method for rarefied flow simulation and the hydrodynamic flow solver for the NS solution have been unified in the UGKWP, where the computational cost of the scheme is on the same order as a particle method in the rarefied regime and as an NS solver in the continuum regime.

Extensive research has been conducted in the past decade under the UGKS framework (Xu & Huang 2010; Huang et al. 2012; Huang et al. 2013; Liu et al. 2016). Besides monatomic gas, the UGKS has been extended to diatomic gas with rotational and vibrational modes (Liu et al. 2014; Wang et al. 2017), the axisymmetric flow (Li et al. 2018), and the flow under the external force field (Xiao et al. 2017). The combination of the UGKS and Lattice Boltzmann Method (LBM) leads to the development of a discrete unified gas-kinetic scheme (DUGKS; Guo et al. 2013; Guo et al. 2015), which has a very wide applicable regime and favourable robustness in comparison with LBM. At the same time, more advanced techniques have been added to the UGKS, such as the moving mesh and velocity space adaptation (Chen et al. 2012), and the implicit and multigrid acceleration (Zhu et al. 2017a; Zhu et al. 2017b; Zhu et al.

2019a). Furthermore, the UGKS methodology has been used to solve other non-equilibrium transport problems. For radiative transfer, the photon transport in the optically thin and thick regions can be treated uniformly under the UGKS framework (Mieussens 2013; Sun et al. 2015a; Sun et al. 2015b; Sun et al. 2017), without the constraint of cell size being less than the photon's mean free path. The solutions from a photon's free transport to the diffusion evolution have been accurately obtained. The schemes have been constructed for multi-group neutron transport (Tan et al. 2019), plasma physics (Liu & Xu 2017), and dispersive multiphase flow (Liu et al. 2019a). The UGKWP is extended to radiative transfer as well, where the ray effect has been totally eliminated in the optically thin region (Li et al. 2020). The scheme can uniformly connect the implicit Monte Carlo (IMC) method in the optically thin region and the diffusion equation solver in the optically thick region (Shi et al. 2020, 2021). In the near future, under the UGKWP framework, the scheme will be further developed in the following areas. In gas dynamics, with the modification of particle collision time according to particle velocity, the UGKWP can recover the DSMC solution accurately in the rarefied flow regime (Xu et al. 2021). A more detailed comparison between the UGKWP and the DSMC will be conducted for non-equilibrium flow. For plasma physics, the UGKWP connects the particle in cell (PIC) method in the collisionless limit and the magneto-hydrodynamic (MHD) solver in the continuum regime. Due to the separate tracking of ions and electrons, all kinds of MHD equations can be recovered by the UGKWP with variations on the Lamor radius and the Knudsen number (Liu & Xu 2021). At the same time, the time step constrained from the speed of light can be removed through the implicit discretization of the Maxwell equations. The UGKWP methodology is expected to solve important multi-scale transport problems in plasma physics, such as magnetic field reconnection and collision-less shock. For multiphase flow, the scheme for the dilute dispersive gas-particle multiphase system will be developed, where the particle is treated by the UGKWP to capture the highly non-equilibrium transport, such as in particle trajectory crossing (PTC), and the gas is simulated by the GKS. The coupling of the GKS and the UGKWP will make the scheme attractable in the study of the gas-particle system. At the same time, the modelling and simulation of non-equilibrium turbulent flow under the UGKS framework may become a point of particular interest. With the multi-scale nature of the UGKWP, besides recovering the diffusive process in the traditional fluid mechanics modelling for turbulence, non-equilibrium transport by including the freely penetrating fluid elements can be incorporated into turbulence modelling as well through the UGKWP methodology. The UGKWP provides an indispensable tool for turbulent flow study, where the particles can be used to track a large number of turbulent eddies and the wave can simulate the large-scale structured flow. In other words, based on a dynamically adaptive distribution of particle and wave, a large-scale coherent flow structure (wave representation) and local fully developed turbulence (particle representation) can be described under a single UGKWP framework. Based on a distribution with a dynamically weighted particle and wave, a unified description can be obtained for both the turbulent and laminar flows. The transition from laminar to turbulent flow can be naturally captured by the emerging of particles through the turbulence Knudsen number as the ratio of the fluid element transport collision time and the numerical time step. In traditional turbulence modelling, only diffusive process or quasi-equilibrium mechanism have been included in Reynolds-Averaged NS (RANS) and Large Eddy Simulation (LES) modelling, starting from the continuum mechanics assumption. Here the UGKWP provides a tool with flexible degrees of freedom and the capability to capture the non-local non-equilibrium transport.

Appendix

In the GKS, UGKS, UGKP, and UGKWP, the moments of an equilibrium state with bounded and unbounded integration limits are needed. The Maxwellian distribution in 3D is

$$g = \rho \left(\frac{\lambda}{\pi}\right)^{\frac{K+3}{2}} e^{-\lambda((u-U)^2 + (v-V)^2 + (w-W)^2 + \xi^2)},$$

where ξ has K degrees of freedom, such as $\xi^2 = \xi_1^2 + \xi_2^2 + \ldots + \xi_K^2$, $K = 0$ for monatomic gas, and $K = 2$ for diatomic gas with two rotational degrees of freedom. By introducing the following notation for the moments of g,

$$\rho \langle \ldots \rangle = \int (\ldots) g \, du dv dw d\xi,$$

where $d\xi = d\xi_1 d\xi_2 \ldots d\xi_K$ and each ξ covers the range of $(-\infty, \infty)$. The general moment formula becomes

$$\langle u^n v^m w^k \xi^l \rangle = \langle u^n \rangle \langle v^m \rangle \langle w^k \rangle \langle \xi^l \rangle,$$

where n, m, and k are positive integers, and l is an even positive integer owing to the symmetrical property of ξ. The moments of $\langle \xi^l \rangle$ are

$$\langle \xi^0 \rangle = 1,$$
$$\langle \xi^2 \rangle = \left(\frac{K}{2\lambda}\right),$$
$$\langle \xi^4 \rangle = \left(\frac{3K}{4\lambda^2} + \frac{K(K-1)}{4\lambda^2}\right),$$

and

$$\langle \xi^{2l} \rangle = \frac{K + 2(l-1)}{2\lambda} \langle \xi^{2(l-1)} \rangle, \text{ for } l = 1, 2, 3, \ldots.$$

The values of $\langle u^n \rangle$ depend on the integration limit. For the limit from $-\infty$ to $+\infty$, the moments are

$$\langle u^0 \rangle = 1,$$
$$\langle u \rangle = U,$$
$$\langle u^{n+2} \rangle = U \langle u^{n+1} \rangle + \frac{n+1}{2\lambda} \langle u^n \rangle, \text{ for } n = 0, 1, 2, 3, \ldots$$

For the moments with limits from 0 to $+\infty$ defined as $\langle\ldots\rangle_{>0}$ and from $-\infty$ to 0 defined as $\langle\ldots\rangle_{<0}$, the moments of u^n in the half-space are

$$\langle u^0\rangle_{>0} = \frac{1}{2}\mathrm{erfc}(-\sqrt{\lambda}U),$$

$$\langle u\rangle_{>0} = U\langle u^0\rangle_{>0} + \frac{1}{2}\frac{e^{-\lambda U^2}}{\sqrt{\pi\lambda}},$$

$$\langle u^{n+2}\rangle_{>0} = U\langle u^{n+1}\rangle_{>0} + \frac{n+1}{2\lambda}\langle u^n\rangle_{>0}, \text{ for } n = 0,1,2,3,\ldots,$$

and

$$\langle u^0\rangle_{<0} = \frac{1}{2}\mathrm{erfc}(\sqrt{\lambda}U),$$

$$\langle u\rangle_{<0} = U\langle u^0\rangle_{<0} - \frac{1}{2}\frac{e^{-\lambda U^2}}{\sqrt{\pi\lambda}},$$

$$\langle u^{n+2}\rangle_{<0} = U\langle u^{n+1}\rangle_{<0} + \frac{n+1}{2\lambda}\langle u^n\rangle_{<0}, \text{ for } n = 0,1,2,3,\ldots.$$

The error function appears in the above solutions. Similarly, the moments $\langle v^m\rangle$ and $\langle w^k\rangle$ can be obtained by changing U to V and W in the above formulations.

In all kinetic schemes presented in this Element, based on derivatives of macroscopic flow variables, such as $\partial\mathbf{W}/\partial x$ or $\partial\mathbf{W}/\partial t$, the corresponding coefficients in the expansion of a Maxwellian distribution function need to be evaluated, such as a and $\overline{A}$ in $\partial g/\partial x = ga$ and $\partial g/\partial t = g\overline{A}$.

With the definition of moments

$$\psi_\alpha = \left(1, u, v, w, \frac{1}{2}(u^2 + v^2 + w^2 + \xi^2)\right)^T \text{ and } d\Xi = dudvdwd\xi,$$

the connection between the derivatives of macroscopic flow variables, such as mass, momentum, energy, and the expansion of the equilibrium distribution function, is

$$\frac{\partial W_\alpha}{\partial x} = \int ga\psi_\alpha d\Xi,$$

where $a = a_\beta\psi_\beta$, such that

$$\mathbf{W} = (\rho, \rho U, \rho V, \rho W, \rho E)^T,$$
$$a = a_1 + a_2 u + a_2 v + a_4 w + a_5 \frac{1}{2}(u^2 + v^2 + w^2 + \xi^2).$$

Then, the above connection gives

$$\frac{\partial W_\alpha}{\partial x} = \int ga\psi_\alpha d\Xi = \int \psi_\alpha\psi_\beta a_\beta g d\Xi,$$

which can be written as

$$\frac{1}{\rho}\frac{\partial W_\alpha}{\partial x} = M_{\alpha\beta}a_\beta,$$

where

$$M_{\alpha\beta} = (1/\rho)\int \psi_\alpha\psi_\beta g d\Xi,$$

and

$$a_\beta = M_{\beta\alpha}^{-1}\frac{1}{\rho}\left(\frac{\partial W_\alpha}{\partial x}\right).$$

The coefficients of a are provided as follows:
Defining

$$A = 2\frac{\partial(\rho E)}{\partial x} - \left(U^2 + V^2 + W^2 + \frac{K+3}{2\lambda}\right)\frac{\partial\rho}{\partial x},$$

$$B = \frac{\partial(\rho U)}{\partial x} - U\frac{\partial\rho}{\partial x},$$

$$C = \frac{\partial(\rho V)}{\partial x} - V\frac{\partial\rho}{\partial x},$$

$$D = \frac{\partial(\rho W)}{\partial x} - W\frac{\partial\rho}{\partial x},$$

the coefficients are

$$a_5 = \frac{4\lambda^2}{(K+3)\rho}(A - 2UB - 2VC - 2WD),$$

$$a_4 = \frac{2\lambda D}{\rho} - Wa_5,$$

$$a_3 = \frac{2\lambda C}{\rho} - Va_5,$$

$$a_2 = \frac{2\lambda B}{\rho} - Ua_5,$$

and

$$a_1 = \frac{1}{\rho}\frac{\partial\rho}{\partial x} - Ua_2 - Va_3 - Wa_4 - \frac{1}{2}\left(U^2 + V^2 + W^2 + \frac{K+3}{2\lambda}\right)a_5.$$

Similar formulations can be obtained for the time derivative of a Maxwellian from the corresponding derivatives of macroscopic flow variables. The spatial derivatives in the y- and z-directions can be obtained similarly.

For the distribution function with discrete particle velocity, the moments will become the summation of all velocity points. The macroscopic variables, stress, and heat flux for monatomic gas $(K = 0)$ can be evaluated as

$$\mathbf{W} = \begin{pmatrix} \rho \\ \rho U \\ \rho V \\ \rho W \\ \rho E \end{pmatrix} = \sum_{i=1}^{nx} \sum_{j=1}^{ny} \sum_{k=1}^{nz} \begin{pmatrix} 1 \\ u_i \\ v_j \\ w_k \\ \frac{1}{2}(u_i^2 + v_j^2 + w_k^2) \end{pmatrix} \omega_i \omega_j \omega_k f_{i,j,k},$$

$$\begin{pmatrix} P_{xx} & P_{xy} & P_{xz} \\ P_{yx} & P_{yy} & P_{yz} \\ P_{zx} & P_{zy} & P_{zz} \end{pmatrix} = \sum_{i=1}^{nx} \sum_{j=1}^{ny} \sum_{k=1}^{nz} \begin{pmatrix} c_i c_i & c_i c_j & c_i c_k \\ c_j c_i & c_j c_j & c_j c_k \\ c_k c_i & c_k c_j & c_k c_k \end{pmatrix} \omega_i \omega_j \omega_k f_{i,j,k},$$

$$\begin{pmatrix} q_x \\ q_y \\ q_z \end{pmatrix} = \frac{1}{2} \sum_{i=1}^{nx} \sum_{j=1}^{ny} \sum_{k=1}^{nz} \begin{pmatrix} c_i \\ c_j \\ c_k \end{pmatrix} (c_i^2 + c_j^2 + c_k^2) \omega_i \omega_j \omega_k f_{i,j,k},$$

where ω_i, ω_j, and ω_k represent the quadrature weights in integrations, respectively, and $c_i = u_i - U$, $c_j = v_j - V$, and $c_k = w_k - W$ represent the peculiar velocity.

In comparison with the moments of a distribution function with a continuous particle velocity space, the use of discrete particle velocity points makes the scheme very expensive in terms of memory requirements and computational costs. The efficiency mainly depends on the number of discrete points used to represent a gas distribution function. The use of the quadrature rule is important for the summations. The order of the quadrature rule will affect the accuracy of macroscopic variables and the satisfaction of conservative constraints for the mass, momentum, and energy.

In general, the high-order quadrature rules are recommended to evaluate the moments.

With the discretization of particle velocity space, a finite domain in the velocity space has been used.

The size of the domain depends on the flow velocity and temperature. Theoretically, the range can be estimated within $\left[-\alpha\sqrt{\frac{1}{2}RT}, \alpha\sqrt{\frac{1}{2}RT}\right]$ and $\alpha \geq 4$. For a given problem, the domain should be large enough to contain all possible contributions of high-speed particles in each direction,

$$|\mathbf{u}|_{min} = \min_{x \in \Omega} \left\{ -|\mathbf{U}| - \alpha \sqrt{\frac{1}{2}RT} \right\}, \quad |\mathbf{u}|_{max} = \max_{x \in \Omega} \left\{ |\mathbf{U}| + \alpha \sqrt{\frac{1}{2}RT} \right\},$$

where the flow velocity $\mathbf{U}$ and temperature T have to be estimated first.

The general numerical quadrature for a function $f(x)$ in a given domain $[a, b]$ can be expressed as

$$\int_a^b f(x)dx = \sum_{i=1}^N \omega_i f(x_i),$$

where x_i are the quadrature points and ω_i are the corresponding quadrature weights of the integration rule. The selection of the quadrature rule depends on the flow problems. Two kinds of quadrature rules are adopted in the kinetic schemes, which are the fourth-order composite Newton-Cotes rule and the Gauss quadrature rule. For the fourth-order Newton-Cotes rule, the domain $[a, b]$ is divided into N sections with equal size $h = (b - a)/N$, the quadrature rule reads

$$\begin{aligned}
\int_a^b f(x)dx &\approx \frac{2h}{45} [7f(x_0) + 32f(x_1) + 12f(x_2) + 32f(x_3) + 7f(x_4) \\
&\quad + 7f(x_4) + \dots + 7f(x_{N-4}) + 7f(x_{N-4}) \\
&\quad + 32f(x_{N-3}) + 12f(x_{N-2}) + 32f(x_{N-1}) + 7f(x_N)] \\
&= \sum_{i=0}^N \omega_i f_i,
\end{aligned}$$

and the quadrature weights ω_i have the following form:

$$\omega_i = \begin{cases}
\dfrac{14h}{45}, & i = 0 \ or \ N; \\[2mm]
\dfrac{28h}{45}, & mod(i, 4) = 0; \\[2mm]
\dfrac{24h}{45}, & mod(i, 4) = 2; \\[2mm]
\dfrac{64h}{45}, & others.
\end{cases}$$

The Newton-Cotes rule can be used for any kind of flow simulation, including the hypersonic or highly non-equilibrium flow. However, the disadvantage of this rule is its enormous memory requirements, due to its equally spaced velocity grid points. For low-speed flow with small temperature variation, the distribution function will be concentrated around zero velocity and be close to an equilibrium state; here, the high-order Gauss-Hermite rule is a preferred choice,

$$\int_{-\infty}^{\infty} f(x)dx = \int_{-\infty}^{\infty} e^{-x^2}[e^{x^2}f(x)]dx \approx \sum_{i=1}^{N} \omega_i e^{x_i^2} f(x_i),$$

where e^{-x^2} is the weighting function, $x_i(i = 1, 2, ..., N)$ are the positive roots of the Hermite polynomial of degree N, and ω_i are the corresponding weights of Gauss-Hermite quadrature, which can be evaluated through

$$\omega_i = \frac{2^{N-1}N!\sqrt{\pi}}{N^2[H_{N-1}(x_i)]^2},$$

where H_i is the ith Gauss-Hermite polynomial (Shizgal 1981). The advantage of the Gauss-Hermite quadrature rule is its relatively high accuracy. However, due to its limited number of quadrature points, this rule cannot be used in cases with high temperatures and high Mach number flow. It should be noted that, for a highly non-equilibrium flow, even when the temperature or fluid velocity is low, the velocity distribution function may become highly erratic and there may be singular points. Thus the limited number of points here may not be sufficient for representing a real distribution, and plausible results may be obtained in high Knudsen number flow simulations. So, in this case, an adaptive mesh in the velocity space is an appropriate choice. The direct use of a particle in the UGKP and the UGKWP provides an optimal adaptation of the particle velocity space.

References

Alexander, F.J., Garcia, A.L., & Alder, B.J. (1998). Cell size dependence of transport coefficients in stochastic particle algorithm. *Phys. Fluids*, 10, 1540–1542.

Aristov, V. (2012). *Direct methods for solving the Boltzmann equation and study of nonequilibrium flows*. Springer Science & Business Media.

Bhatnagar, P.L., Gross, E.P., & Krook, M. (1954). A model for collision processes in gases. I. Small amplitude processes in charged and neutral one-component systems. *Phys. Rev.* 94 (3), 511.

Bird, G. (1970). Aspects of the structure of strong shock waves. *Phys. Fluids* 13, 1172–1177.

Bird, G. (1994). *Molecular gas dynamics and the direct simulation of gas flows*. Oxford Science Publications.

Burt, J., Josyula, E., Deschenes, T., & Boyd, I. (2011). Evaluation of a hybrid Boltzmann-continuum method for high-speed nonequilibrium flows. *J. Thermophys. Heat Trans.* 25, 500–515.

Chacon, L., Chen, G., Knoll, D., Newman, C., Park, H., Taitano, W., Willert, J., & Womeldorff, G. (2017). Multiscale high-order/low-order (HOLO) algorithms and applications. *J. Comput. Phys.* 330, 21–45.

Chapman, S., Cowling, T.G., & Burnett, D. (1990). *The mathematical theory of non-uniform gases: an account of the kinetic theory of viscosity, thermal conduction and diffusion in gases*. Cambridge University Press.

Chen, S., & Doolen, G. (1998). Lattice Boltzmann method for fluid flows. *Ann. Rev. Fluid Mech.* 30, 329–364.

Chen, S.Z., Xu, K., & Cai, Q. (2015). A comparison and unification of ellipsoidal statistical and Shakhov BGK. *Adv. Appl. Math. Mech.* 7, 245–266.

Chen, S.Z., Xu, K., Lee, C.B., & Cai, Q. (2012). A unified gas kinetic scheme with moving mesh and velocity space adaptation. *J. Comput. Phys.* 231 (20), 6643–6664.

Chen, S.Z, Zhang, C., Zhu, L., & Guo, Z. (2017). A unified implicit scheme for kinetic model equations. part I. memory reduction technique. *Science Bull.* 62 (2), 119–129.

Chen, Y.P., Zhu, Y.J., & Xu, K. (2020). A three-dimensional unified gas-kinetic wave-particle solver for flow computation in all regimes. *Phys. Fluids* 32, 096108.

Chou, S., & Baganoff, D. (1997). Kinetic flux-vector splitting for the Navier-Stokes equations. *J. Comput. Phys.* 130, 217–230.

Chu, C.K. (1965). Kinetic-theoretic description of the formation of a shock wave. *Phys. Fluids* 8 (1), 12–22.

Degond, P., Dimarco, G., & Mieussens, L. (2010). A multiscale kinetic–fluid solver with dynamic localization of kinetic effects. *J. Comput. Phys.* 229 (13), 4907–4933.

Degond, P., Liu, J., & Mieussens, L. (2006). Macroscopic fluid models with localized kinetic upscale effects. *Multiscale Model. Simul.* 5, 940–979.

Deshpande, S. (1986). A second order accurate, kinetic-theory based method for inviscid compressible flows. NASA Langley Tech. Paper No. 2613.

Dimarco, G., & Pareschi, L. (2013). Asymptotic preserving implicit-explicit Runge-Kutta methods for nonlinear kinetic equations, *SIAM J. Numer. Anal.* 51, 1064–1087.

Eu, B. (2016). *Kinetic theory of nonequilibrium ensembles, irreversible thermodynamics, and generalized hydrodynamics: Volume 1. Nonrelativistic theories*. Springer.

Fei, F., Zhang, J., Li, J., & Liu, Z. (2020). A unified stochastic particle Bhatnagar-Gross-Krook method for multiscale gas flows. *J. Comput. Phys.* 400, 108972.

Filbet, F., & Jin, S. (2010). A class of asymptotic-preserving schemes for kinetic equations and related problems with stiff sources. *J. Comput. Phys.* 229 (20), 7625–7648.

Gallis, M., & Torczynski, J. (2000). The application of the BGK model in particle simulations. In *34th Thermophysics Conference*, 2360.

Ghia, U., Ghia, K., & Shin, C. (1982). High-Resolutions for incompressible flow using the Navier-Stokes equations and a multigrid method. *J. Comput. Phys.* 48, 387–411.

Gorji, M.H., & Jenny, P. (2015). Fokker-Planck-DSMC algorithm for simulations of rarefied gas flows. *J. Comput. Phys.* 287, 110–129.

Grad, H. (1949). On the kinetic theory of rarefied gases. *Commun. Pure Appl. Math.* 2, 325.

Gu, X., & Emerson, D. (2009). A high-order moment approach for capturing non-equilibrium phenomena in the transition regime. *J. Fluid Mech.* 636, 177–216.

Guo, Z., Li, J., & Xu, K. (2020). On unified preserving properties of kinetic schemes. arXiv: 1909.04923v4.

Guo, Z., Wang, R., & Xu, K. (2015). Discrete unified gas kinetic scheme for all Knudsen number flows. II. Thermal compressible case. *Phys. Rev. E* 91 (3), 033313.

Guo, Z., Xu, K., & Wang, R. (2013). Discrete unified gas kinetic scheme for all Knudsen number flows: Low-speed isothermal case. *Phys. Rev. E* 88 (3), 033305.

Huang, J.C., Xu, K., & Yu, P. (2012). A unified gas-kinetic scheme for continuum and rarefied flows II: Multidimensional cases. *Comm. Comput. Phys.* 12 (3), 662–690.

Huang, J.C., Xu, K., & Yu, P. (2013). A unified gas-kinetic scheme for continuum and rarefied flows III: Microflow simulations. *Comm. Comput. Phys.* 14 (5), 1147–1173.

Jenny, P., Torrilhon, M., & Heinz, S. (2010). A solution algorithm for the fluid dynamic equations based on a stochastic model for molecular motion. *J. Comput. Phys.* 229 (4) 1077–1098.

Ji, X., Pan, L., Shyy, W., & Xu, K. (2018b). A compact fourth-order gas-kinetic scheme for the Euler and Navier–Stokes equations. *J. Comput. Phys.* 372, 446–472.

Ji, X., Zhao, F., Shyy, W., & Xu, K. (2018a). A family of high-order gas-kinetic schemes and its comparison with Riemann solver based high-order methods. *J. Comput. Phys.* 356, 150–173.

Ji, X., Zhao, F., Shyy, W., & Xu, K. (2020). A HWENO reconstruction based high-order compact gas-kinetic scheme on unstructured meshes. *J. Comput. Phys.* 410, 109367.

Jiang, D.W., Mao, M.L., & Deng, X.G. (2019a). An implicit parallel UGKS solver for flows covering various regimes. *Adv. Aerodynamics* 1:8, https://doi.org/10.1186/s42774-019-0008-5.

Jiang, Z., Zhao, W., Yuan, Z., Chen, W., & Myong, R. (2019b). Computation of hypersonic flows over flying configurations using a nonlinear constitutive model. *AIAA J.* 57 (12), 5252–5268.

Jin, C., & Xu, K. (2007). A unified moving grid gas-kinetic method in Eulerian space for viscous flow computation. *J. Comput. Phys.* 222, 155–175.

Jin, S. (1999). Efficient asymptotic-preserving (AP) schemes for some multiscale kinetic equations. *SIAM J. Sci. Comp.* 21 (2), 441–454.

John, B., Gu, X., & Emerson, D. (2011). Effects of incomplete surface accommodation on non-equilibrium heat transfer in cavity flow: A parallel DSMC study. *Comput. Fluids* 45, 197–201.

Kolobov, V., Arslanbekov, R., Aristov, V., Frolova, A., & Zabelok, S. (2007). Unified solver for rarefied and continuum flows with adaptive mesh and algorithm refinement. *J. Comput. Phys.* 223, 589–608.

Kumar, G., Girimaji, S., & Kerimo, J. (2013). WENO-enhanced gas-kinetic scheme for direct simulations of compressible transition and turbulence. *J. Comput. Phys.* 234, 499–523.

Larsen, A., Morel, J., & Miller, W. (1987). Asymptotic solutions of numerical transport problems in optically thick, diffusive regimes. *J. Comput. Phys.* 69 (2),283–324.

Lele, S. (1992). Compact finite difference schemes with spectral-like resolution. *J. Comput. Phys.* 103, 16–42.

Levermore, C. (1996). Moment closure hierarchies for kinetic theories. *J. Statistical Phys.* 83, 1021.

Li, J., & Du, Z. (2016). A two-stage fourth order time-accurate discretization for Lax–Wendroff type flow solvers I: Hyperbolic conservation laws. *SIAM J. Sci. Comput.* 38 (5), A3046–A3069.

Li, Q., Xu, K., & Fu, S. (2010). A high-order gas-kinetic Navier-Stokes solver. *J. Comput. Phys.* 229, 6715–6731.

Li, S., Li, Q., Fu, S., & Xu, K. (2018). A unified gas-kinetic scheme for axisymmetric flow in all Knudsen number regimes. *J. Comput. Phys.* 366, 144–169.

Li, W., Liu, C., Zhu, Y., Zhang, J., & Xu, K. (2020). Unified gas-kinetic wave-particle methods III: Multiscale photon transport. *J. Comput. Phys.* 408, 109280.

Li, Z.H., & Zhang, H.X. (2004). Study on gas kinetic unified algorithm for flows from rarefied transition to continuum. *J. Comput. Phys.* 193 (2), 708–738.

Liu, C. (2016). Unified gas-kinetic scheme for the study of multi-scale flows. PhD thesis, Hong Kong University of Science and Technology.

Liu, C., Wang, Z., & Xu, K. (2019a). A unified gas-kinetic scheme for continuum and rarefied flows VI: Dilute disperse gas-particle multiphase system. *J. Comput. Phys.* 386, 264–295.

Liu, C., & Xu, K. (2017). A unified gas kinetic scheme for continuum and rarefied flows V: Multiscale and multicomponent plasma transport. *Comm. Comput. Phys.* 22 (5), 1175–1223.

Liu, C., & Xu, K. (2020a). A unified gas-kinetic scheme for micro flow simulation based on linearized kinetic equation. *Adv. Aerodynamics* 2:21, https://doi.org/10.1186/s42774-020-00045-8.

Liu, C., & Xu, K. (2021). Unified gas-kinetic wave-particle methods IV: Multi-species gas mixture and plasma transport. *Adv. Aerodynamics* 3:9, https://doi.org/10.1186/s42774-021-00062-1

Liu, C., Xu, K., Sun, Q., & Cai, Q. (2016). A unified gas-kinetic scheme for continuum and rarefied flows IV: Full Boltzmann and model equations. *J. Comput. Phys.* 314, 305–40.

Liu, C., Zhou, G., Shyy, W., & Xu, K. (2019b). Limitation principle for computational fluid dynamics. *Shock Waves* 29:1083–1102.

Liu, C., Zhu, Y., & Xu, K. (2020). Unified gas-kinetic wave-particle methods I: Continuum and rarefied gas flow. *J. Comput. Phys.* 401, 108977.

Liu, S., Yu, P., Xu, K., & Zhong, C. (2014). Unified gas-kinetic scheme for diatomic molecular simulations in all flow regimes. *J. Comput. Phys.* 259, 96–113.

Luo, J., & Xu, K. (2013). A high-order multidimensional gas-kinetic scheme for hydrodynamic equations. *Science China, Technological Sciences* 56, 2370–2384.

Luo, J., Xuan, L., & Xu, K. (2013). Comparison of fifth-order WENO scheme and WENO-gas-kinetic scheme for inviscid and viscous flow simulation. *Commun. Comput. Phys.* 14, 599–620.

Macrossan, M.N. (2001). A particle simulation method for the BGK equation. In *AIP Conference Proceedings*, Vol. 585, AIP, 426–433.

Mieussens, L. (2000). Discrete velocity model and implicit scheme for the BGK equation of rarefied gas dynamics. *Math. Models Methods Applied Sci.* 10 (8), 1121–1149.

Mieussens, L. (2013). On the asymptotic preserving property of the unified gas kinetic scheme for the diffusion limit of linear kinetic model. *J. Comput. Phys.* 253, 138–156.

Mouhot, C., & Pareschi, L. (2006). Fast algorithms for computing the Boltzmann collision operator. *Math. Comp.* 75 (256), 1833–1852.

Myong, R. (2001). A computational method for Eu's generalized hydrodynamic equations of rarefied and microscale gas dynamics. *J. Comput. Phys.* 168, 47–72.

Ohwada, T., Adachi, R., Xu, K., & Luo, J. (2013). On the remedy against shock anomalies in kinetic schemes. *J. Comput. Phys.* 255, 106–129.

Ohwada, T., & Kobayashi, S. (2004). Management of discontinuous reconstruction in kinetic schemes. *J. Comput. Phys.* 197, 116–138.

Oran, E., Oh, C., & Cybyk, B. (1998). Direct simulation Monte Carlo: Recent advances and applications. *Ann. Rev. Fluid Mech.* 30 (1), 403–441.

Pan, L., & Xu, K. (2016). A third-order compact gas-kinetic scheme on unstructured meshes for compressible Navier–Stokes solutions. *J. Comput. Phys.* 318, 327–348.

Pan, L., Xu, K., Li, Q., & Li, J. (2016). An efficient and accurate two-stage fourth-order gas-kinetic scheme for the Euler and Navier–Stokes equations. *J. Comput. Phys.* 326, 197–221.

Pareschi, L., & Russo, G. (2000). Asymptotic preserving Monte Carlo methods for the Boltzmann equation. *Trans. Theory Stat. Phys.* 29 (3–5) 415–430.

Perthame, B. (1992). Second-order Boltzmann schemes for compressible Euler equations in one and two space dimensions. *SIAM J. Numer. Anal.* 29, 1–19.

Pieraccini, S., & Puppo, G. (2007). Implicit-explicit schemes for BGK kinetic equations. *J. Sci. Comput.* 32, 1–28.

Pullin, D. (1980). Direct simulation methods for compressible inviscid ideal gas flow. *J. Comput. Phys.* 34, 231–244.

Ren, W., Liu, H., & Jin, S. (2014). An asymptotic-preserving Monte Carlo method for the Boltzmann equation. *J. Comput. Phys.* 276, 380–404.

Ren, X.D., Xu, K., & Shyy, W. (2016). A multidimensional high-order DG-ALE method based on gas-kinetic theory with application to oscillating bodies. *J. Comput. Phys.* 316, 700–720.

Schwartzentruber, T.E., & Boyd, I. (2007). A molecular particle-continuum method for hypersonic nonequilibrium gas flows. *J. Comput. Phys.* 225, 1159–1174.

Schwartzentruber, T.E., Scalabrin, L.C., & Boyd, I. (2008). Hybrid particle-continuum simulations of nonequilibrium hypersonic blunt-body flowfields. *J Thermophys. Heat Trans.* 22 (1), 29–37.

Shakhov, E. (1968). Generalization of the Krook kinetic relaxation equation. *Fluid Dyn.* 3 (5), 95–96.

Shi, Y., Song, P., & Sun, W. (2020). An asymptotic preserving unified gas kinetic particle method for radiative transfer equations. *J. Comput. Phys.* 420, 109687.

Shi, Y., Sun, W., Li, L., & Song, P. (2021). An improved unified gas kinetic particle method for radiative transfer equations. *J. Quant. Spectrosc. Radiat. Transfer* 261,107428.

Shizgal, B. (1981). A Gaussian quadrature procedure for use in the solution of the Boltzmann equation and related problems. *J. Comput. Phys.* 41, 309–328.

Struchtrup, H. (2005). *Macroscopic transport equations for rarefied gas flows: Approximation methods in kinetic theory.* Springer.

Struchtrup, H., & Torrihon, M. (2003). Regularization of Grad's 13 moment equations: Derivation and linear analysis. *Phys. Fluids* 15, 2668.

Su, W., Zhu, L., Wang, P., Zhang, Y., & Wu, L. (2020). Can we find steady-state solutions to multiscale rarefied gas flows within dozens of iterations? *J. Comput. Phys.* 407, 109245.

Succi, S. (2015). Lattice Boltzmann 2038. *EPL* 109 (5), 50001.

Sun, W., Jiang, S., & Xu, K. (2015a). An asymptotic preserving unified gas kinetic scheme for gray radiative transfer equations. *J. Comput. Phys.* 285, 265–279.

Sun, W., Jiang, S., & Xu, K. (2017). A multidimensional unified gas-kinetic scheme for radiative transfer equations on unstructured mesh. *J. Comput. Phys.* 351, 455–472.

Sun, W., Jiang, S., Xu, K., & Li, S. (2015b). An asymptotic preserving unified gas kinetic scheme for frequency dependent radiative transfer equations. *J. Comput. Phys.* 302, 222–238.

Tan, S., Sun, W., Wei, J., & Ni, G. (2019). A parallel unified gas kinetic scheme for three-dimensional multi-group neutron transport. *J. Comput. Phys.* 391, 37–58.

Tcheremissine, F. (2005). Direct numerical solution of the Boltzmann equation. In *AIP Conference Proceedings*, Vol. 762, AIP, 677–685.

Tiwari, S. (1998). Coupling of the Boltzmann and Euler equations with automatic domain decomposition. *J. Comput. Phys.* 144, 710–726.

Tumuklu, O., Li, Z., & Levin, D.A. (2016). Particle ellipsoidal statistical Bhatnagar-Gross-Krook approach for simulation of hypersonic shocks. *AIAA J.* 3701–3716.

van Leer, B. (1977). Towards the ultimate conservative difference scheme IV: a new approach to numerical convection. *J. Comput. Phys.* 23, 276–299.

van Leer, B. (1979). Towards the ultimate conservative difference scheme V: A second order sequel to Godunov's method. *J. Comput. Phys.* 32, 101–136.

Vincenti, W., & Kruger, C. (1965). *Introduction to physical gas dynamics*. Krieger.

Wagner, W. (1992). A convergence proof for Bird's direct simulation Monte Carlo method for the Boltzmann equation. *J. Statistical Phys.* 66, 1011–1044.

Wang, Z., Yan, H., Li, Q., & Xu, K. (2017). Unified gas-kinetic scheme for diatomic molecular flow with translational, rotational, and vibrational modes. *J. Comput. Phys.* 350, 237–259.

Wijesinghe, H.S., & Hadjiconstantinou, N. (2004). Three-dimensional hybrid continuum-atomistic simulations for multiscale hydrodynamics. *J. Fluid Engineering* 126, 768–777.

Woodward, P., & Colella, P. (1984). Numerical simulations of two-dimensional fluid flow with strong shocks. *J. Comput. Phys.* 54, 115–173.

Wu, L., Reese, J.M., & Zhang, Y. (2014). Solving the Boltzmann equation deterministically by the fast spectral method: application to gas microflows. *J. Fluid Mech.* 746, 53–84.

Wu, L., White, C., Scanlon, T.J., Reese, J.M., & Zhang, Y. (2013). Deterministic numerical solutions of the Boltzmann equation using the fast spectral method. *J. Comput. Phys.* 250, 27–52.

Wu, L., Zhang, J., Reese, J.M., & Zhang, Y. (2015). A fast spectral method for the Boltzmann equation for monatomic gas mixtures. *J. Comput. Phys.* 298, 602–621.

Xiao, T., Cai, Q., & Xu, K. (2017). A well-balanced unified gas-kinetic scheme for multiscale flow transport under gravitational field. *J. Comput. Phys.* 332, 475–491.

Xiao, T., Liu, C., Xu, K., & Cai, Q. (2020). A velocity-space adaptive unified gas kinetic scheme for continuum and rarefied flows. *J. Comput. Phys.* 415, 109535.

Xu, K. (2001). A gas-kinetic BGK scheme for the Navier–Stokes equations and its connection with artificial dissipation and Godunov method. *J. Comput. Phys.* 171 (1), 289–335.

Xu, K. (2015). *Direct modelling for computational fluid dynamics: Construction and application of unified gas-kinetic scheme.* World Scientific.

Xu, K., He, X., & Cai, C. (2008). Multiple temperature kinetic model and gas-kinetic method for hypersonic nonequilibrium flow computations. *J. Comput. Phys.* 227, 6779–6794.

Xu, K., & Huang, J.C. (2010). A unified gas-kinetic scheme for continuum and rarefied flows. *J. Comput. Phys.* 229 (20), 7747–7764.

Xu, K., & Huang, J.C. (2011). An improved unified gas-kinetic scheme and the study of shock structures. IMA *J. App. Math.* 76, 698–711.

Xu, K., & Li, Z. (2001). Dissipative mechanism in Godunov-type schemes. *Int. J. Numer. Meth. Fluids* 37, 1–22.

Xu, K., & Liu, C. (2017). A paradigm for modelling and computation of gas dynamics. *Phys. Fluids* 29, 026101.

Xu, K., Mao, M., & Tang, L. (2005). A multidimensional gas-kinetic BGK scheme for hypersonic viscous flow. *J. Comput. Phys.* 203, 405–421.

Xu, X., Chen, Y., Liu, C., Li, Z., & Xu, K. (2020). Unified gas-kinetic wave-particle methods V: Diatomic molecular flow. arXiv:2010.07195v1 [physics.comp-ph] 14 Oct.

Xu, X., Chen, Y., & Xu, K. (2021). Modelling and computation of non-equilibrium gas dynamics: Beyond single relaxation time kinetic models. *Phys. Fluids* 33, 011703; doi: 10.1063/5.0036203.

Yang, J., & Huang, J. (1995). Rarefied flow computations using nonlinear model Boltzmann equations. *J. Comput. Phys.* 120 (2), 323–339.

Yang, L., Shu, C., Yang, W., & Wu, J. (2018). An implicit scheme with memory reduction technique for steady state solutions of DVBE in all flow regimes. *Phys. Fluids* 30 (4), 040901.

Yu, P. (2013). A unified gas kinetic scheme for all Knudsen number flows. PhD thesis, Hong Kong University of Science and Technology.

Zhao, F., Ji, X., Shyy, W., & Xu, K. (2019). Compact higher-order gas-kinetic schemes with spectral-like resolution for compressible flow simulations. *Adv. Aerodynamics* 1:13, https://doi.org/10.1186/s42774-019-0015-6.

Zhao, F., Ji, X., Shyy, W., & Xu, K. (2020a). An acoustic and shock wave capturing compact high-order gas-kinetic scheme with spectral-like resolution. *Int. J. Comput. Fluid. Dyn.* https://doi.org/10.1080/10618562.2020.1821879.

Zhao, F., Ji, X., Shyy, W., & Xu, K. (2020b). Compact high-order gas-kinetic scheme on unstructured mesh for acoustic and shock wave computations. arXiv:2010.05717v2 [math.NA] 19 Oct 2020.

Zhong, X.L., MacCormack, R.W., & Chapman, D. (1993). Stabilization of the Burnett equations and application to hypersonic flows. *AIAA J.* 31 (1), 1036.

Zhu, L., Guo, Z., & Xu, K. (2016). Discrete unified gas kinetic scheme on unstructured meshes. *Comp. Fluids* 127, 211–225.

Zhu, L., Pi, X., Su, W., Li, Z.H., Zhang, Y., & Wu, L. (2021). General synthetic iteration scheme for non-linear gas kinetic simulation of multi-scale rarefied gas flows. *J. Comput. Phys.* 430, 110091.

Zhu, Y.J., Liu, C., Zhong, C.W., & Xu, K. (2019b). Unified gas-kinetic wave-particle methods II: Multiscale simulation on unstructured mesh. *Phys. Fluids* 31, 067105.

Zhu, Y.J., Zhong, C., & Xu, K. (2017a). Implicit unified gas-kinetic scheme for steady state solution in all flow regimes. *J. Comput. Phys.* 315, 16–38.

Zhu, Y.J., Zhong, C., & Xu, K. (2017b). Unified gas-kinetic scheme with multigrid convergence for rarefied flow study. *Phys. Fluids* 29, 096102.

Zhu, Y.J., Zhong, C., & Xu, K. (2019a). An implicit unified gas-kinetic scheme for unsteady flow in all Knudsen regimes. *J. Comput. Phys.* 386, 190–217.

Zhu, Y.J., Zhong, C.W., & Xu, K. (2020). Ray effect in rarefied flow simulation. *J. Comput. Phys.* 422, 109751.

Acknowledgement

The author would like to thank all his students and collaborators who have worked on the unified gas-kinetic schemes over the past decade. Thanks also go to funding support from the Hong Kong Research Grant Council and National Natural Science Foundation of China.

Cambridge Elements ≡

Elements of Aerospace Engineering

Vigor Yang

Georgia Institute of Technology

Vigor Yang is the William R. T. Oakes Professor in the Daniel Guggenheim School of Aerospace Engineering at Georgia Tech. He is a member of the US National Academy of Engineering and a Fellow of the American Institute of Aeronautics and Astronautics (AIAA), American Society of Mechanical Engineers (ASME), Royal Aeronautical Society (RAeS), and Combustion Institute (CI). He is currently a co-editor of the Cambridge University Press Aerospace Series and co-editor of the book *Gas Turbine Emissions* (Cambridge University Press, 2013).

Wei Shyy

Hong Kong University of Science and Technology

Wei Shyy is President of Hong Kong University of Science and Technology and a Chair Professor of Mechanical and Aerospace Engineering. He is a fellow of the American Institute of Aeronautics and Astronautics (AIAA) and the American Society of Mechanical Engineers (ASME). He is currently a co-editor of the Cambridge University Press Aerospace Series, co-author of *Introduction to Flapping Wing Aerodynamics* (Cambridge University Press, 2013) and co-editor in chief of *Encyclopedia of Aerospace Engineering*, a major reference work published by Wiley-Blackwell.

About the Series

An innovative new series focusing on emerging and well-established research areas in aerospace engineering, including advanced aeromechanics, advanced structures and materials, aerospace autonomy, cyber-physical security, electric/hybrid aircraft, deep space exploration, green aerospace, hypersonics, space propulsion, and urban and regional air mobility. Elements will also cover interdisciplinary topics that will drive innovation and future product development, such as system software, and data science and artificial intelligence.

Cambridge Elements ≡

Elements of Aerospace Engineering

Elements in the Series

Distinct Aerodynamics of Insect-Scale Flight
Csaba Hefler, Chang-kwon Kang, Huihe Qiu and Wei Shyy

A Unified Computational Fluid Dynamics Framework from Rarefied to Continuum Regimes
Kun Xu

A full series listing is available at: www.cambridge.org/EASE

Printed in the United States
by Baker & Taylor Publisher Services

Printed in the United States
by Baker & Taylor Publisher Services